Understanding Integrated Circuits

MURRAY P. ROSENTHAL
Manager, Engineering Services
RCA Global Communications, Inc.

HAYDEN BOOK COMPANY, INC.
Rochelle Park, New Jersey

Library of Congress Cataloging in Publication Data

Rosenthal, Murray P
Understanding integrated circuits.

Includes index.
SUMMARY: An introduction to the operation, fabrication, and application of linear and digital integrated circuits, with a brief synopsis of digital logic and solid-state physics.
1. Integrated circuits—Amateurs' manuals.
[1. Integrated circuits—Amateurs' manuals.
2. Electronics] I. Title.
TK9965.R67 621.381'73 75-31879
ISBN 0-8104-5526-9

Printed in the United States of America

2	3	4	5	6	7	8	9	PRINTING
	77	78	79	80	81	82	83	YEAR

Preface

This book has been written to provide an understanding of the fundamental principles involved in the design and application of integrated circuits. It was prepared primarily to assist technical students, hobbyists, technicians, and engineers, who already have a working knowledge of fundamental electronics but who wish to broaden their knowledge of integrated electronics.

The material is divided into six chapters. Chapter 1 introduces the reader to the various broad categories of integrated circuits. Chapter 2 offers a brief introduction to digital logic symbols and circuits, and their functions. Chapter 3 reviews the basics of semiconductor physics, such as materials, current flow, junctions, biasing, and devices. Chapter 4 covers IC fabrication and packaging. Chapter 5 describes several simple and complex applications, both linear and digital. Chapter 6 outlines some practical hints on IC and general solid-state testing and troubleshooting. A glossary of some of the more common IC and digital terminology has also been included, following Chapter 6.

MURRAY P. ROSENTHAL

New York City

Acknowledgment

The information provided herein is based upon the work, both published and unpublished, of a great many companies and individuals. However, the author wishes to express special thanks to the following for their contributions to and cooperation in the preparation of this book: ALCO Electronic Products; AMP, Inc.; AP Products, Inc.; Cambridge Thermionic Corp. (CAMBION); Dionics, Inc.; *Electronic Design*; Electronic Products; Fairchild Semiconductor; General Automation, Inc.; General Instrument Corp.; Helipot Div. of Beckman Instruments; Hewlett-Packard, Santa Clara Div.; IBM Corp.; Inselek Corp.; Micro Electronic Systems, Inc.; MOSTEK Corp.; Motorola Semiconductor; National Semiconductor Corp.; Philips Research Labs; RCA Corp., especially Ms. E. McElwee and Messrs. A. Acampora, N. DiSanti, J. T. Frankle, D. Hampel, and N. Kotsolios; Signetics Corp.; Simpson Electric Co.; Spectrum Dynamics; Teledyne Semiconductor; Texas Instruments, Inc.; Toshiba R & D Center; Triplett Corp.; Ungar Div. of Eldon Industries; and, above all, my wife, Rita, who kept the coffee hot and my temper cool under some very trying conditions.

M. P. R.

Contents

1
Introduction

The Integrated Circuit

Introduction

The diode, first of the solid-state devices is a small two-element device that permits current flow in only one direction, and has no gain. Then came the transistor; also small, also solid-state, but with at least three elements and with the ability to amplify current while dissipating little heat. And now, we have the integrated circuit (IC). The IC shown in Fig. 1-1 performs within its tiny framework all the functions of the discrete transistorized module which, in turn, performs all the functions of the tube module shown.

This newest technology in the field of electronics, integrated circuits, is not really new at all, it has been around for more than a decade. However, the last five years have seen ICs virtually take over the electronics industry. For example, where we once thought the minicomputer, the desk-top version of the big floor models, was the last word, Hewlett-Packard has produced a *hand-held* calculator with a *memory*. Users can prepare and edit programs on the new HP-65 Calculator, then store them on tiny magnetic cards that can be used over and over again or erased and rerecorded upon.

But just what is an IC? It's not simply a printed circuit as we know it, i.e., a replacement for hard wiring. Yet an IC *is* a printed circuit, in that its components are printed out from a diagram of the desired circuit. One difference is that, while the printed circuit is simply a printed wiring layout, the IC is a combination of the wiring layout and semiconductor physics. In simpler terms, an integrated circuit is one in which the component parts (resistors, capacitors, transistors, etc.) are semiconductor substances that work in accordance with the laws of semiconductor physics, and are interconnected by microscopic printed-circuit wiring to perform a particular function.

Fig. 1-1. The IC contains as many functions as its discrete predecessors. (Courtesy Texas Instruments, Inc.)

A second difference is that, while a discrete component, e.g., a resistor, can be removed from a printed circuit without harm to the latter, ICs are inseparable and irreparable, and must be removed and replaced in entirety. Thus we arrive at a final definition of an IC: a microphotograph reproduced within semiconductor materials in order to obtain the functions of an entire electronic circuit within the smallest possible space.

Advantages of Integrated Circuits

The question can be asked, "Why bother with this new technology? Discrete components perform pretty well in their own right." There are several reasons for so bothering. First, and most obvious, is the tremendous savings possible in space and weight. Consider, for example, a medium-sized digital computer. In 1955, using tubes, the space it would have occupied would have been equivalent to several large rooms. In 1965, using discrete transistors, one room might have been enough. Today, the same computer using ICs fits on a desk top, and briefcase computers are already a reality.

A second advantage of ICs lies in their greater reliability. This is partially due to the fact that, except for hybrid ICs, all interconnections, along with the various components, are fabricated in the initial manufacturing process. Also contributing to their reliability is the use of highly refined manufacturing and testing techniques during the production process.

A third advantage is that ICs require much less power to operate than do their discrete equivalents. This, in turn, results in cheaper operation.

A fourth advantage lies in extended frequency response. As there are no lengthy wire connections in an IC amplifier, and since the size of each component is decreased, the frequency response of the circuit is extremely good.

Taken all together, the stated advantages make it readily apparent why IC technology has advanced so rapidly.

Types of Integration

There are three types of integration. First, simple integration consisting of less than ten elements, such as the diode quad on a single "chip" shown in Fig. 1-2. Second, medium-scale integration (MSI),

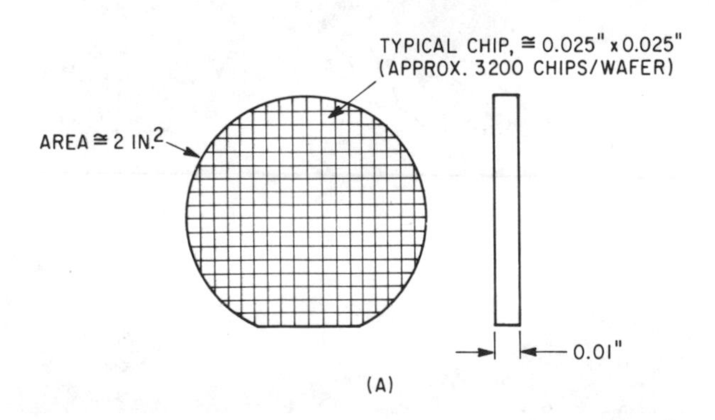

Fig. 1-2. Simple integration. (A) Typical silicon wafer, exaggerated for clarity.

Fig. 1-2. Simple integration. (B) D1914 Diode Quad; chip thickness = 6 mils + 1 mil. (Courtesy Dionics, Inc.)

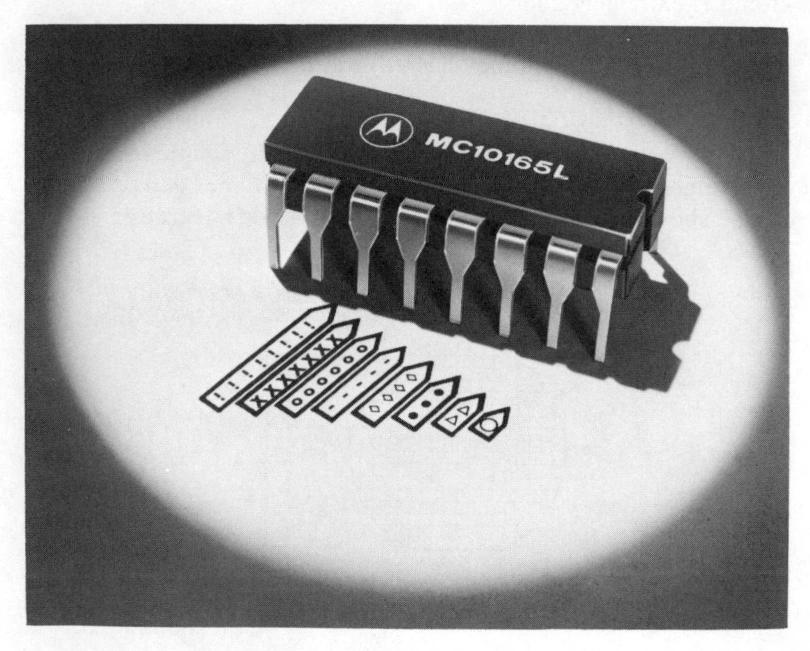

Fig. 1-3. Medium-scale integration (MSI) priority encoder. (Courtesy Motorola Semiconductor)

Fig. 1-4. Large-scale integration (LSI), 1024-Bit Dynamic Random-Access Memory. (Courtesy MOSTEK Corp.)

used largely for computer circuits, generally consisting of between 10 and 100 "gates" (i.e., circuits as complex as a logic gate) on a single chip, as shown in Fig. 1-3. Third, large-scale integration (LSI), which by implication, consists of a chip containing 100 or more gates, as typified by the IC shown in Fig. 1-4.

Monolithic Integrated Circuits

In a monolithic integrated circuit, such as that shown in Fig. 1-5, active electronic elements, i.e., diodes and transistors, are fabricated upon or within a single semiconductor substrate, usually silicon, together with passive elements, i.e., resistors and capacitors, with at least one element being formed within the semiconductor substrate itself. The term "monolithic" derives from the word "monolith," meaning single stone. Here, the "stone" is usually a single chip of

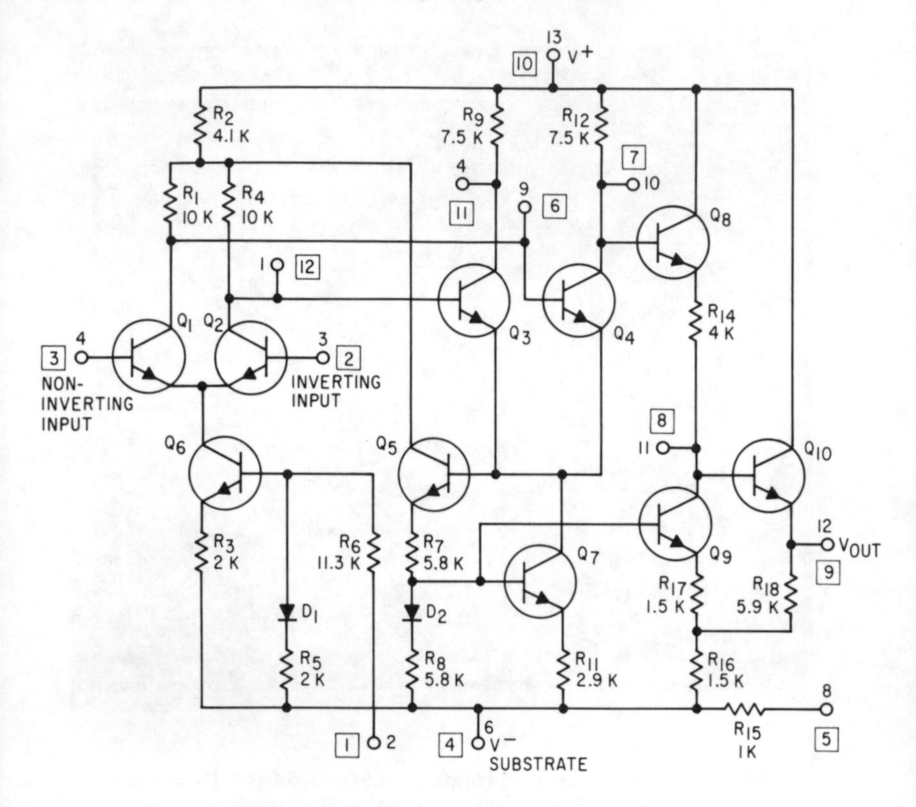

Fig. 1-5. Schematic diagram of a monolithic IC operational amplifier. (Courtesy RCA Corp.)

silicon whose dimensions are on the order of 0.0254 cm (centimeter) thick x 0.064 cm square (cm^2).

Thin-Film Integrated Circuits

The general classification of thin-film ICs applies to a structure that begins with a substrate fabricated from an insulating material such as glass, glazed alumina, or glazed ceramic. On this substrate are deposited a pattern of the passive circuit elements as well as an interconnecting metalization pattern that connects both these elements and the active elements of the circuit. The latter are added to the thin film separately.

Thin-film depositions are performed in a vacuum chamber (hence the term "vacuum deposition") and are usually only 100-150A

(Angstroms) thick. As of this writing, only conductors, resistors, capacitor dielectrics, and crossover insulators are feasible commercially, but work on producing active devices is progressing rapidly.

In addition to the straightforward thin-film technique is the process termed "compatible thin-film." The term compatible indicates that the process used to deposit the thin film on top of the insulating layer has no effect on the IC beneath the insulating layer. The process begins with the basic monolithic IC. On top of this IC is then deposited an insulating layer of silicon dioxide (SiO_2). The thin film is then deposited on top of this dioxide insulation. The thin-film-to-monolithic interconnections are made through pre-etched holes (windows) in the dioxide layer.

Thick-Film Integrated Circuits

The distinction between thick and thin-film ICs is not merely that of dimension, but rather in the fabrication process. Thick films are formed by printing or screening (somewhat like silk-screening) a pattern of resistive, conductive, or dielectric material onto an insulating substrate, which is usually a ceramic consisting of 96 percent Alumina ($Al_2 O_3$). (Alumina is being replaced by soda-lime glass, which is ordinary window glass, as the substrate for electro-optical display devices, such as readouts.)

Thick-film circuits are generally combined with semiconductor circuits or other devices to give complete hybrid circuits. But they can also be formed into ICs and resistive IC networks.

Hybrid Integrated Circuits

In airborne, mobile, or portable applications, power dissipation, size, and weight are generally quantities desirable to minimize. ICs offer one alternative to the attainment of this end. However, one can't always afford the customizing of MSI or LSI for the particular application, nor does the particular IC always exist for the particular application. Thus, the need for hybrids.

Hybrid, or multi-chip integrated circuits are complete electronic circuits in which a multiplicity of separately manufactured items are arrayed on a suitable passive substrate, such as ceramic, and interconnected by metalization patterns and/or very fine wires. In short, a hybrid may be defined as any combination of two or more of the following: discrete components, an active-substrate IC, and a passive-substrate IC. At present, individual discrete transistors are used in most hybrids as the active elements, with the passive components being either discrete resistors and capacitors or thin- or thick-film resistors and capacitors. Lastly, a hybrid may also consist of one or more monolithic IC chips housed in a single package, together with assorted auxiliary components. Motorola, for example, has developed what they term MSHI—medium-scale hybrid integration—which they cram into one 30-lead, 1 in. x 1 in. flatpack. Six of these flatpacks, containing both silicon-gate CMOS logic chips (ICs) and a few discrete components, took the place of 34 TTL ICs, 47 discrete diodes, resistors, and capacitors, and a delay line—in less than half the space and with less than half the power requirement.

Linear and Digital Integrated Circuits

Briefly, linear ICs are those concerned with amplification of an input signal, i.e., they act upon the signal itself. Digital ICs are those

Fig. 1-6. Typical IC packages. (A) 12-pin, TO-5 round style. (B) 24-pin flatpack. (C) 14-pin ceramic MSI dual inline (DIP). (D) 14-pin plastic dual inline (DIP). (Courtesy RCA Corp.)

concerned with the "state" of a signal, i.e., they act dependent upon the *level* of the signal. By level is meant zero voltage, or above or below any predetermined reference voltage.

Linear ICs are generally used in consumer equipment such as TV receivers, stereos, and AM-FM receivers. Also, because of the special nature of linear-digital interfaces, semiconductor manufacturers have devised a variety of linear ICs (op amps, comparators, sense amplifiers, line drivers, etc.) peculiarly suited to such applications. On the other hand, digital ICs are presently confined mostly to computer or instrumentation circuits, but are also finding increased application in automotive electronics.

Integrated Circuit Packaging

Conventional integrated circuits are currently packaged in three distinct configurations: Round, flatpack, and dual inline (DIP) as shown in Fig. 1-6. The round style, which is similar in looks to a typical TO-5 transistor, is a glass-metal package, and comes in either 8-, 10-, or 12-lead configurations. The flatpack is generally made of ceramic, and comes in 10-, 14-, 16-, 24-, and 32-lead configurations. The DIP is either ceramic or plastic, and comes in 10-, 14-, 16-, 24-,

Fig. 1-7. Typical logic module comprising several DIP ICs and providing a random-access memory of up to 4,000 words. (Courtesy Cambridge Thermionic Corp.)

28-, 36-, or 40-lead configurations. The variation in shapes is due simply to the need for design engineers to be able to fit the ICs physically into the space they have available. For example, Fig. 1-7 illustrates how several DIP ICs have been crammed into a typical logic module to provide a random-access memory of up to 4,000 words.

A recent innovation is the "leadless" IC. The concept is similar to the old vacuum-tube/socket arrangement: the IC is plugged into the permanently-mounted socket, with contact being established between the side-mounted IC metalization and the metal fingers of the socket. The advantages of the leadless IC include greater chip area available for added functions, easy removal of faulty ICs, and use of the socket's comparatively large fingers as test points.

2
Digital Logic

Introduction

Automated or programmed devices using IC logic circuits become more common daily. The reader whose background lies mostly with conventional (analog type) circuits may have difficulty understanding digital diagrams filled with strange symbols. Most people with an electronics background are trained to use schematic diagrams which require consideration of each individual component and its contribution toward circuit operation. In today's logic circuitry, the point of interest is shifted upward. Rather than considering each individual piece, entire circuits are supplied in individual packages. It is not necessary to know the exact circuit configuration of the package because it is encapsulated. Consider a typical NAND gate, the Fairchild 914. Within the IC are six transistors and numerous associated resistors. The only access to these components is through the eight IC pins. Therefore, signal tracing the circuitry through the IC is impossible. It is only necessary to understand the relationship between the input and output signals.

This chapter will help the reader to approach digital concepts. It begins with a brief review of the concepts of decimal and binary number systems. Next, the rules of Boolean algebra and their application to the field of digital logic are offered. It then presents the application of the algebra to the design, simplification, and understanding of these circuits.

Binary and Decimal Numbers

Digital equipment may be broken down into hundreds (or thousands) of switching devices. A switch has two stable conditions: "on" and "off." When examining devices containing many switches, the decimal system is impractical. As the switch is a two-state device, a counting or numbering system based on the value 2 is convenient. Such a numbering system is called the binary numbering system. Although probably unfamiliar to the average reader, the binary numbering system is logical and easily understood.

In the decimal number system 10 symbols are used: 0, 1, 2, 3, 4, 5, 6, 7, 8, and 9. Someone counting pennies, for example, and writing down the count, would write it as shown above, 0 through 9. For the tenth penny, however, he has run out of symbols, and he starts again with 0 and places a 1 to the left of the 0, indicating the count has reached 10 one time. The next count is 11, indicating one 10 + one 1 = 11. When the count reaches 20, note that the right-hand column begins with 0 again, but this time a 2 is written to the left of the 0. This indicates that the count has gone to 10 a total of two times. The symbol 73 indicates seven 10s + three 1s. Note that at the count 99, the symbols have again been exhausted, so he repeats the change that occurred at 10 and writes 100, indicating one 100 + zero 10s + zero 1s.

The change points are even powers of 10, which are indicated as $10^0 = 1$; $10^1 = 10$; $10^2 = 100$; $10^3 = 1000$; etc. In a written number such as 10,758 we can determine the various powers of 10 that the number represents by the position of the written numbers, i.e., $1 \times 10^4 + 0 \times 10^3 + 7 \times 10^2 + 5 \times 10^1 + 8 \times 10^0$.

The binary numbering system, by comparison, uses only two symbols, the first two symbols of the Arabic numbering system: 0 and 1. To see how binary counting works, again assume someone is counting pennies and has to write the count in binary form. He begins by writing 0, indicating that he has not yet counted. He then counts the first penny and writes 1. He now has on his paper 0, 1. When he counts the second penny, what then? Since there are only two symbols in the Binary system, he therefore resorts to the same method used in the decimal system: he writes a 0 and places a 1 to the left of it, indicating he has counted to 2 one time. At the count of 3 he writes 11, indicating one 2 + one 1 = 3. At the count of 4 he is again out of symbols, so he writes 100, indicating one 4 + zero 2s + zero 1s. At the count of 5 he writes 101, indicating one 4 + zero 2s + one 1. He continues until he reaches the count of 7, where he writes 111. Again he has used all symbols in all columns, so for the count of 8 he writes 1000, indicating one 8 + zero 4s + zero 2s + zero 1s. Table 2-1 shows the binary count along with the same count in decimal form.

Comparing the same number in binary and decimal forms shows that the binary form is clumsy in that it takes many more digits to express a number. Note from Table 2-1 that 32 requires 6 digits to

Table 2-1. Comparison of binary and decimal numbers.

Decimal	Binary	Decimal	Binary
0	0	17	10001
1	1	18	10010
2	10	19	10011
3	11	20	10100
4	100	21	10101
5	101	22	10110
6	110	23	10111
7	111	24	11000
8	1000	25	11001
9	1001	26	11010
10	1010	27	11011
11	1011	28	11100
12	1100	29	11101
13	1101	30	11110
14	1110	31	11111
15	1111	32	100000
16	10000		

express in binary form. Why, then, does digital equipment use the binary system? Simply because electronic devices and decimal counting are not very compatible. Although circuits can be constructed to use base-10 values, the circuitry would be quite complex and would involve the use of 10 different voltage levels. On the other hand, active electronic devices can operate as two-level devices, i.e., switches, and two voltage-level circuits are easy to make. Also, such devices can be made to switch at rates of millions of times a second. Thus, it is simplicity and speed which make the use of the binary system practical in electronics.

Digital devices use data and instructions in binary form. We, however, use decimal numbers and alphabetic letters. Therefore, various codes have been devised to facilitate communication via digital equipment. These codes are formed by taking groups of bits (BInary digiTs) and assigning each unique combination a particular letter, symbol, or decimal number. There are many codes in existence, only one of which is considered here, as it is probably the most common of all.

Table 2-2. 8, 4, 2, 1 BCD code.

Decimal	BCD	Decimal	BCD
0	0000	5	0101
1	0001	6	0110
2	0010	7	0111
3	0011	8	1000
4	0100	9	1001

The simplest code to understand is the binary-coded decimal, BCD. The BCD code uses 4 bits per character (group of bits) and a weighting scheme of 8, 4, 2, and 1. Each character has the decimal value that the 4 bits represent. The code is illustrated in Table 2-2. Note that the decimal equivalent is simply the binary number expressed in decimal form.

A 4-bit number can have values from 0 through 15. Ordinarily, however, only enough combinations are used in the BCD code to express all 10 decimal symbols. To express decimal numbers greater than 9, a separate 4-bit group is used for each number. For example, 82 is expressed as 1000 0010 in BCD. Note that the BCD requires many bits to express the equivalent decimal number.

Boolean Algebra

The reader's understanding of digital circuits requires an understanding of a different form of algebra from that taught in high school: Boolean algebra. Boolean algebra differs from conventional algebra in that it uses the binary numbering system. Boolean algebra contains methods specially adaptable to digital circuitry and facilitates the design of such circuitry. Conventional algebra, on the other hand, may needlessly complicate digital circuit design.

Boolean algebra is a method of manipulating deductive logic. It recognizes only two possible values for a statement. A statement is either entirely true or entirely false; there are no halfway conditions. A statement which is not true must therefore be false. These premises

Fig. 2-1 Illustrating Boolean logic. (A) "Open" or "false" switch. (B) "Closed" or "true" switch.

allow the algebra to be used to represent the conditions found in electrical switching circuitry. Consider the switch of Fig. 2-1. The switch is either open or closed. It has no other possible conditions. By applying the basic premise of Boolean algebra we can define the closed switch as a "true" condition and the open switch as a "false" condition. The switch, when not open, must be closed. If not closed it is open. This parallels the Boolean logic. The closed condition could be called the true state and the open condition a false state, without ambiguity. By defining the conditions of a two-state device in Boolean terms, the symbology of the algebra becomes usable. Note the switch of Fig. 2-1 can also itself represent various electronic elements such as transistors and diodes operated in switched modes.

Inverter, OR, AND, and Exclusive-OR

Inverter Function

The inverter, sometimes called a NOT circuit, is a one-input, one-output device. Figure 2-2 represents the logic symbol for an inverter and illustrates a positive input G and an inverted output G (overscore denotes negation).

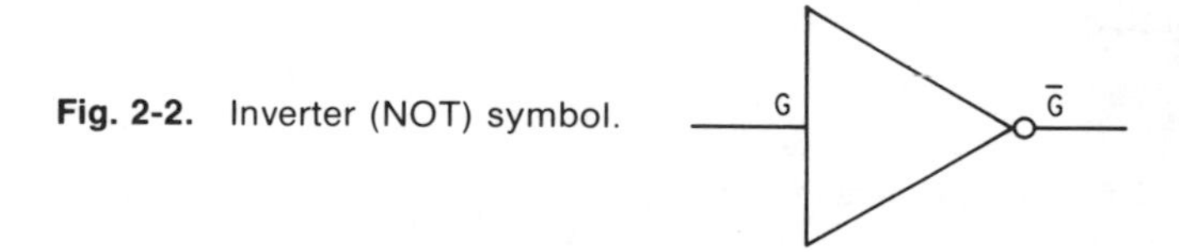

Fig. 2-2. Inverter (NOT) symbol.

OR Function

Consider Fig. 2-3 as representing an OR gate. When either or both switches are closed, the output will be true. Only when both switches are open will the output be false.

Reading aloud, the reader should state, "A or B" equals hi. The Boolean equation is represented as follows:

$$A + B = 1$$

where "+" in Boolean algebra is read as OR and 1 is read as true (hi).

Literally, the equation states, "If either A or B (or both) is true, then the output is true." A table listing all possible combinations of A and B is required to examine this equation properly. Such a table is easily constructed as shown in Fig. 2-4.

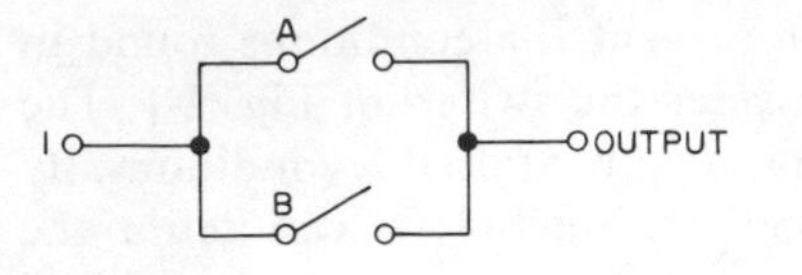

Fig. 2-3. The OR gate function.

A	B	Output
0	0	0
1	0	1
0	1	1
1	1	1

Fig. 2-4. OR-function truth table.

The table shows that there are four possible combinations for the variables A and B. It follows that in a different equation with three variables there would be eight combinations and with four variables, sixteen combinations, etc.

The foregoing describes a Boolean OR function. For the purpose of simplifying diagrams wherever a circuit appears which could perform an OR function, the schematic may be replaced by the OR gate symbol as shown in Fig. 2-5.

There are cases when negative logic is required. To accomplish this, the absolute voltage level assigned to logic 1 is set at a negative level with respect to the voltage level assigned to logic zero. The exact voltage levels could be any values desired, but if the more negative of the two values set represents logic 1, then the logic is negative; and if the more positive of the two represents logic 0, then the logic is also negative.

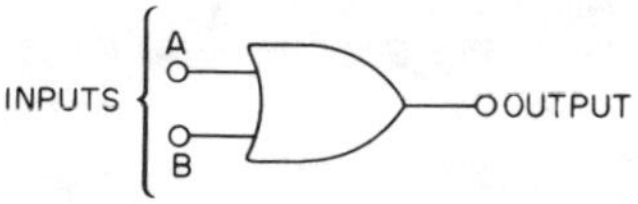

Fig. 2-5. The OR gate logic symbol.

AND Function

Another important logic function is the one performed by the AND gate. Figure 2-6A represents a simple circuit of two switches A and B connected in series and is used here to explain the AND concept.

By referring to this figure it can be seen that the way the diagram is presented, the circuit will not conduct, i.e., no output (0). Closing both switches A and B completes the circuit thus allowing for the input to be felt at the output (1). The Boolean equation for an AND function can therefore be represented as follows:

$$A \cdot B \quad 1$$

where "·" in Boolean algebra means AND.

A	*B*	*Output*
0	0	0
1	0	0
0	1	0
1	1	1

(B)

Fig. 2-6. The AND function and its truth table.

Literally, the equation states, "When A and B are both true, then and only then is the output also true." A truth table listing all possible combinations of A and B is shown in Fig. 2-6B. The schematic diagram and its equivalent logic symbol are shown in Figs. 2-7A and 2-7B, respectively.

Figure 2-7A shows that when the inputs A and B are both present, the npn transistors will both conduct, allowing for a high output. If on the other hand the input to the base of either transistor is low, the output would also be low (at ground potential).

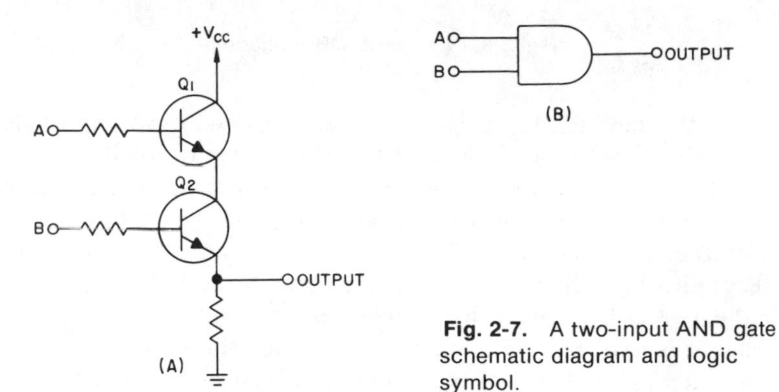

Fig. 2-7. A two-input AND gate schematic diagram and logic symbol.

Exclusive-OR Function

Suppose now the problem is to implement the following logical statement. "A room with two doors is to have a lamp installed with switches accessible to each door, either one of which can turn the lamp either on or off." Let A represent the switch at one door, B the switch at the other door, and L stand for the lamp. We first construct the truth table (Fig. 2-8A). Although the choice is entirely arbitrary, assume that when a switch is closed it has a logical value of 1, and when opened it has the logical value of 0.

When A and B are both zero, the lamp will be off—we assign this condition a logical value of zero. In order for either switch to control the lamp, if A or B changes states, the lamp must go on. Therefore,

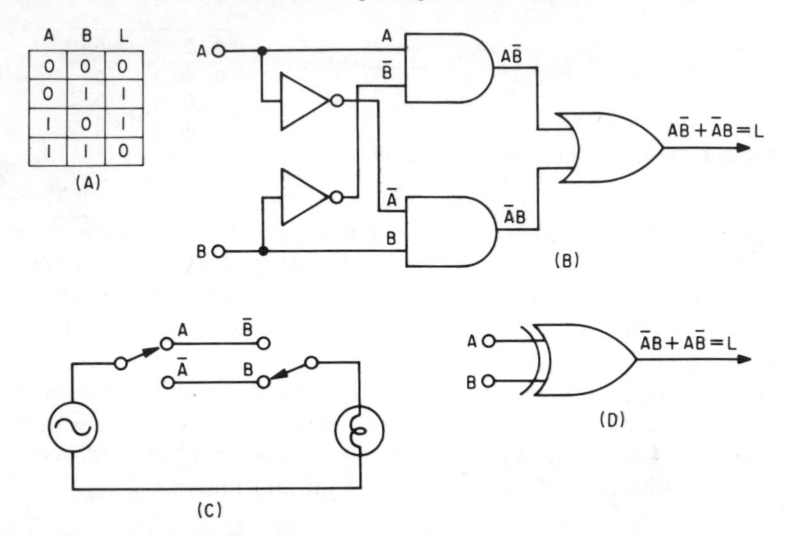

Fig. 2-8. Exclusive-OR functions.

when A is 0 and B is 1, L must be 1. When A is 1 and B is 0, L must also be 1. Finally, when A and B are both 1, the lamp must be off. The next problem is to write a Boolean algebra expression for the truth table.

L is to be 1 for two possible conditions; when A is 0 and B is 1, and when A is 1 and B is 0. When two quantities are ANDed together, they must be both equal to 1 for the result to be 1. In this case we indicate A as 0, however, its complement, $\overline{A}$, would have a value of 1 when A is zero. Therefore, we write $\overline{A}B$. Similarly, we write $A\overline{B}$. L is 1 for either one OR the other combination. The complete equation is $\overline{A}B + A\overline{B} = L$.

Next, we implement the equation in a logic diagram. Wherever two terms are ANDed together we use the symbol for an AND gate; wherever the two terms are ORed together, we use the symbol for an OR gate. The complete diagram is shown in Fig. 2-8B. Inverters are used to generate the negated values of any one of the variables when it is required.

Figure 2-8C shows the final circuit using switches to implement the logic functions. By the use of single-pole double-throw switches we generate both the true and negated value for a particular variable.

Reexamining Fig. 2-8B, note that L is true if either A or B is true, but not when both are true. This particular combination is so useful that it has been given a special name and a special symbol. The implementing circuit is called an "Exclusive-OR" gate. The symbol is shown in Fig. 2-8D which diagrams the lamp problem using the Exclusive OR. The problem could be diagrammed as in Figs. 2-8B or 2-8C; however, Fig. 2-8D is the more convenient.

NAND Gate, NOR Gate, and Flip-Flop

A Boolean algebra function may be implemented by using any common switching device, such as a transistor. When operated in the common-emitter mode, the transistor acts as an inverter. Thus, when a transistor is used to implement a logic function, the output of the transistor, if taken at the collector, represents the inversion of the input operation. This means that when constructing an AND gate using a transistor, the output frequently represents the inversion of an AND gate.

Figure 2-9A shows a two-input digital gate, consisting of two transistors, Q1 and Q2. In a negative logic system, when inputs A and B are true they are 0V. In this condition, both transistors are off, and the output level is +5V, the false level. Examination of the truth table, (Fig. 2-9B) for an AND gate shows that the output is 0 or false for any combination of inputs A and B except when both are true. Referring to Fig. 2-9A again, if both transistors are off, which occurs only when A and B are true (0V), Y is false (+5V). For all other conditions, one or the other or both of the transistors is on because its base is at the false or positive level. With either transistor on, Y is qual to 0V, the true level. The Y column is the inverse of an AND gate output.

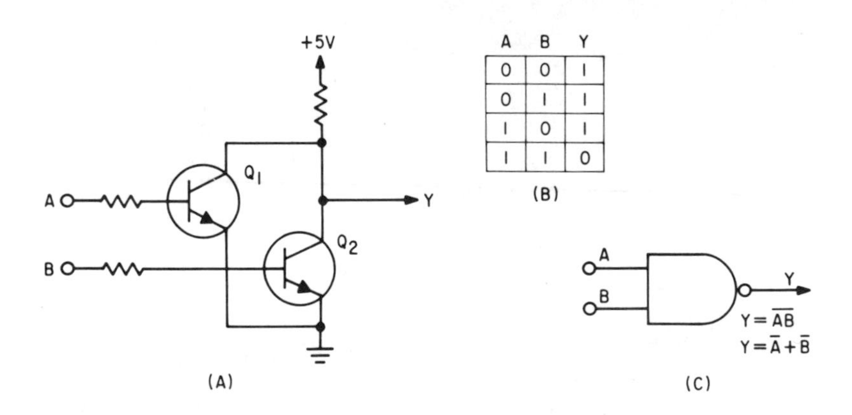

A	B	Y
0	0	1
0	1	1
1	0	1
1	1	0

Fig. 2-9. A two-input NAND gate.

NAND Gate

The type of gate known as a negated AND gate, is shortened to NAND gate. The symbol for a NAND gate appears in Fig. 2-9C. The basic shape of the gate identifies it as an AND function. The circle at the output of the gate means that the signal is logically inverted at the point where the circle appears. The Boolean algebra expression for the NAND gate is written as $Y = \overline{AB}$, or $Y = \overline{A} + \overline{B}$. Note that in one form of the equation, the OR function is indicated.

NOR Gate

Figure 2-10A shows a digital logic circuit using two transistors. This figure represents a NOR gate. It operates as follows:

When either input A or input B or both inputs are true (hi), the appropriate transistor is forward biased. This forward bias turns on the transistor and allows a path for the V_{CC} to flow to ground. Thus, a false (lo) output.

Only when both inputs are false do the transistors get cut off. With both the transistors cut off, V_{CC} is felt at the output. The truth table for the NOR function is shown in Fig. 2-10B. As shown in Fig. 2-10C, the basic logic symbol for a NOR gate is very similar to that of an OR gate, with the difference being the addition of the circle at the output. This circle indicates logic inversion.

A Boolean equation for a NOR gate could be presented as follows:

$$\overline{A + B} = 0 \qquad or \qquad \overline{A} \,.\, \overline{B} = 0$$

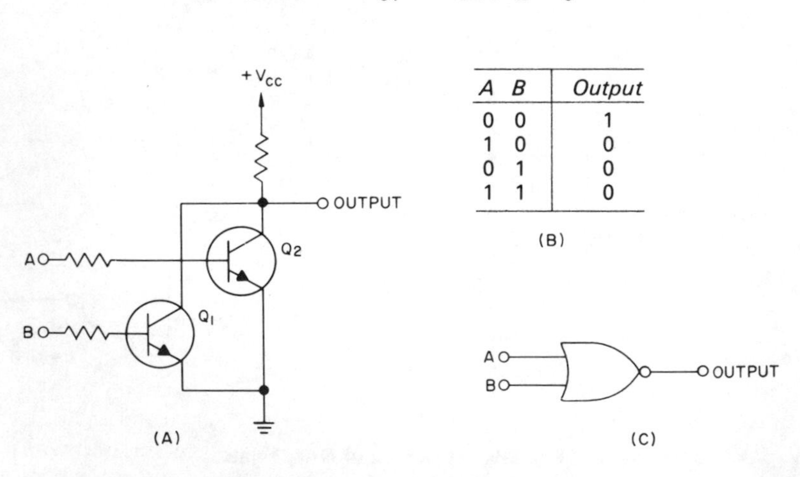

A	*B*	*Output*
0	0	1
1	0	0
0	1	0
1	1	0

Fig. 2-10. A two-input NOR gate.

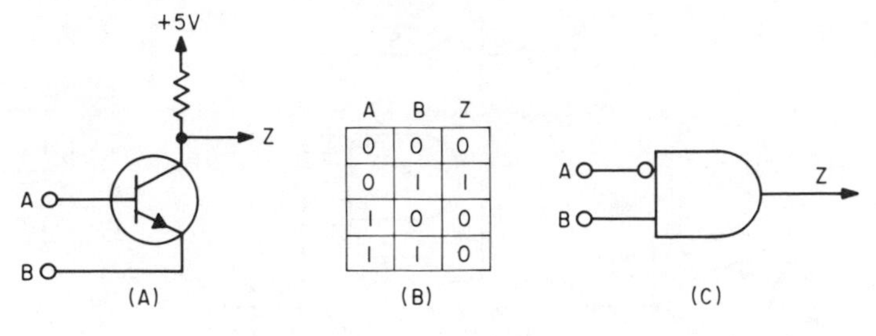

Fig. 2-11. An inhibited gate.

Inhibitor

Figure 2-11A illustrates a transistor whose input terminals are the emitter and the base. In this case, input A must be "high" and input B must be "low" for the transistor to be on. The inputs required have opposite levels. The truth table for such a gate in a negative logic environment is shown in Fig. 2-11B. The output is zero (or false) except when A is zero and B is one. The logic symbol is shown in Fig. 2-11C. The equation for such a gate is $Z = \overline{A}B$. Occasionally such a gate is used where it is desired to hold one input false and disable the entire gate. Here, if input B is held false, the input at A has no control over the gate. In this sense, then, input B is termed an inhibiting input and the gate is often called an "inhibitor," a special form of AND or NAND gate with mixed-logic inputs.

Flip-Flop

The logic operations discussed to this point are essentially single action in nature. A set of logic signals is applied to a set of logic decision devices which proceed to generate a single result. Many digital systems require nothing more. Others, perhaps the majority, require a series of such operations in sequence. To provide a sequential action, a device is required which has a memory; one which will remember the results of a logic operation for later use. A counting circuit is an example. When counting from 1 to 10 the counter must remember how many units have already been counted. At the fourth count, to realize that this is the fourth count, the counting circuits must remember that three prior counts have been made.

A simple way of making a memory device is by using an OR gate, as shown in Fig. 2-12A. OR gate E feeds back a portion of its output to input B. Assume that X initially equals 0. If X is 0, then A must be

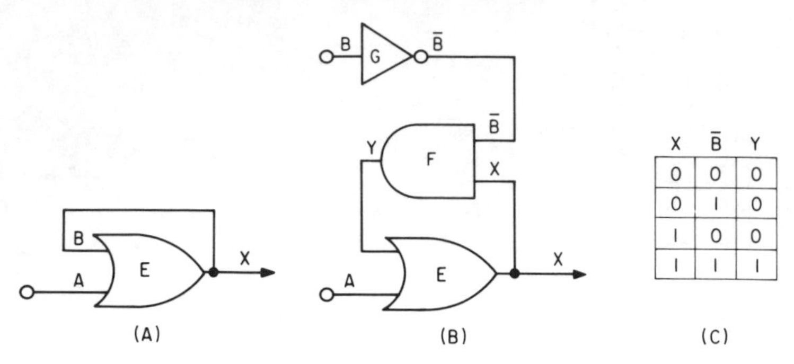

X	B̄	Y
0	0	0
0	1	0
1	0	0
1	1	1

Fig. 2-12. Evolving a flip-flop.

0. If A is now made equal to 1, then X must equal 1. If X equals 1, input B receives the 1 and, regardless of whether A returns to 0, the output remains at 1. The circuit as shown is impractical because there is no way to force X back to 0. However, the gate can be forced back to 0 if we can find some means of breaking the feedback loop. Figure 2-12B illustrates a method of so doing; now the feedback loop contains AND gate F, which has inputs $\overline{B}$ and X and output Y. The truth table for gate F is shown in Fig. 2-12C. From the table we see that if $\overline{B}$ is 1, Y equals X. If $\overline{B}$ is 0, however, the output is 0. Thus, by making $\overline{B}$ equal to 0 we can force OR gate E back to 0, if A is also 0 at the same time. (The logic equation for this circuit is $Y = \overline{B} \cdot X$.) Looking now at the overall circuit, we see that a 1 input at A will force output X to 1; whilc a 1 at input B, inverted by gate G, forces X to 0.

We now have an electronic equivalent to a toggle switch. Flip the switch one way to on, flip the switch the other way to off. Similarly, make B equal to 1 and X flips to 0. Make A equal to 1 and X flips to 1. This leads to the definition for a binary memory unit. Such a binary unit has two control inputs. A true level at one input forces the device output to 1. The other input having a true input forces the output to the opposite state.

The circuit of Fig. 2-12B is bistable. If input A goes to 1, X goes to 1. If input B goes to 1, X goes to 0. Thus, we have the so-called flip-flop (FF) action.

Toggle Flip-Flop A commonly encountered type of FF using transistors is shown in Fig. 2-13. A trigger pulse introduced at A is differentiated and then conducted via diodes D1 and D2 to the bases of Q1 and Q2. D1 and D2 polarize the pulse so that only its negative edge appears at the bases. The transistors are turned off by a negative

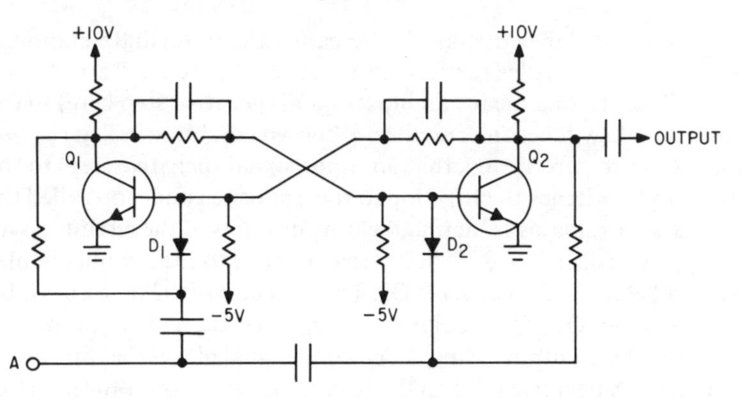

Fig. 2-13. Typical toggle flip-flop.

edge. Any time a trigger pulse is coupled into input A, the flip-flop will change state. Since this is analogous to toggle-switch action, the circuit is called a toggle flip-flop (T FF). The T FF has a major shortcoming: The state of the FF *after* a trigger is applied cannot be accurately known unless the *present* state is known.

Set-Reset Flip-Flop Another common FF is the "Set-Reset" (RS FF). A representative transistor RS FF is shown in Fig. 2-14A. This is similiar to the T FF, except that there are two input terminals. This circuit is predictable for three of four input conditions.

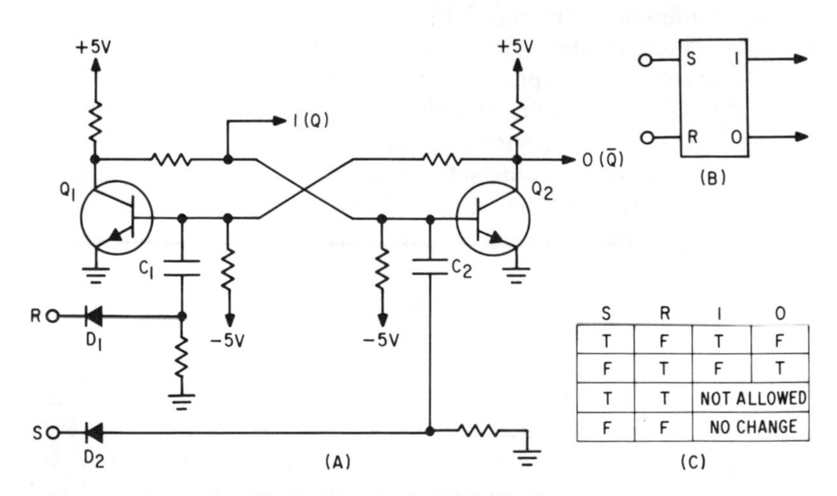

S	R	1	0
T	F	T	F
F	T	F	T
T	T	NOT ALLOWED	
F	F	NO CHANGE	

Fig. 2-14. Typical RS flip-flop.

The inputs labeled R and S are called the Reset and Set inputs, respectively. Two rules for RS flip-flops are "Set to one" and "Reset to zero." Set to one means an input signal (negative step here) to the Set terminal switches the circuit to a known condition called the *one state*. Reset to zero means that an input signal (negative step) to the Reset input switches the flip-flop to the opposite condition called the *zero state*. It remains to define the output states of the circuit. As an example, consider Fig. 2-14A. A negative edge to the S input couples through D2 and C2 to turn off Q2. The collector of Q2 then goes false and turns on Q1. Q1's collector then goes true. This output is, therefore, the 1 output. Thus, a Set input signal places the flip-flop in the 1 state. A negative edge to the R input switches the 1 output to 0.

The collector of Q2 always logically complements the 1 output, and is called the 0 output. Figure 2-14B shows the logic diagram symbol for a typical RS flip-flop. The box symbol shows two inputs on the left, two outputs on the right. The truth table (Fig. 2-14C) shows that all input conditions are covered except when S and R inputs receive simultaneous negative edges. The next state of the FF cannot be predicted and is ambiguous. Thus, for the RS flip-flop, simultaneous R and S inputs are commonly called "not allowed" or "forbidden" combinations. The RS flip-flop is used in logic situations which do not include the possibility of simultaneous Set and Reset inputs.

Binary Element One of the more sophisticated forms of flip-flop circuit is illustrated in Fig. 2-15. This is called a binary element because it includes the functions of sampling or clocking the input signal and storing the input temporarily while the setting action takes place. The binary element also blocks any further input during the setting period. This temporary storage is very important because it holds the information being set into the flip-flop while all the flip-flops in the system are changing. If the temporary storage were not provided, the input signal which originates in another flip-flop could change during the setting period, causing the wrong value to be entered. In general, if simple flip-flops are used, it takes two flip-flops to store and handle one bit of information. One binary element alone can handle one bit of information.

Temporary storage is provided by the clock pulse input capacitors. When the clock signal goes positive, there will be zero charge stored on these capacitors if both S_C and R_C are positive. If either S_C or R_C is negative, the corresponding capacitor will be

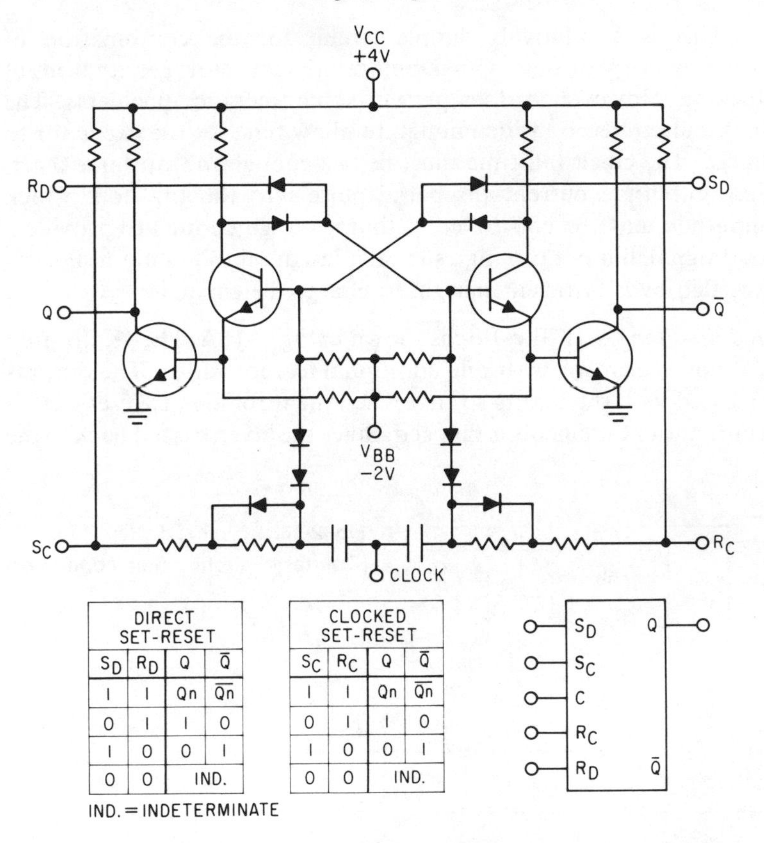

DIRECT SET-RESET			
S_D	R_D	Q	$\bar{Q}$
1	1	Qn	$\overline{Qn}$
0	1	1	0
1	0	0	1
0	0	IND.	

CLOCKED SET-RESET			
S_C	R_C	Q	$\bar{Q}$
1	1	Qn	$\overline{Qn}$
0	1	1	0
1	0	0	1
0	0	IND.	

IND. = INDETERMINATE

Fig. 2-15. Signetics SE124 RS binary element. (Courtesy Signetics)

charged negatively through the resistor and charging diode. When the clock pulse goes negative, the signal input end of the capacitor will be driven more negative than the setting signal. The very negative signal will be coupled up to the flip-flop through the two bias diodes, delivering a negative signal which turns off that side of the flip-flop.

The setting action occurs during the fall-time of the clock pulse. The charging diode is back-biased when the capacitor drives below the input signal level. Thus, any further input is shut off during the clock fall-time. The energy to turn the flip-flop off is stored in the capacitor during this time when the input is blocked. This is called a “steered circuit.” It is sometimes called “diode steering” and also called “pulse pedestal gating.”

This is a relatively simple circuit for the combination of functions it provides: Clocking, temporary storage, and input blocking. However, it does present some tolerance problems. The clock pulse must be wide enough to allow time for the capacitor to charge. The clock fall-time must be fast enough to transfer a short, high amplitude current discharge pulse into the flip-flop. Clock amplitude must be controlled so that it is high enough to provide a good signal, but not too high so that a low amplitude 1 input signal is exceeded by a sufficient margin to charge the capacitor.

JK Flip-Flop A JK flip-flop is shown in Fig. 2-16A. The JK flip-flop is a binary element with one additional feature added. The outputs are gated with the inputs so that when the forbidden two set signals occur, one of the signals is blocked. Since the inverse is fed back to the

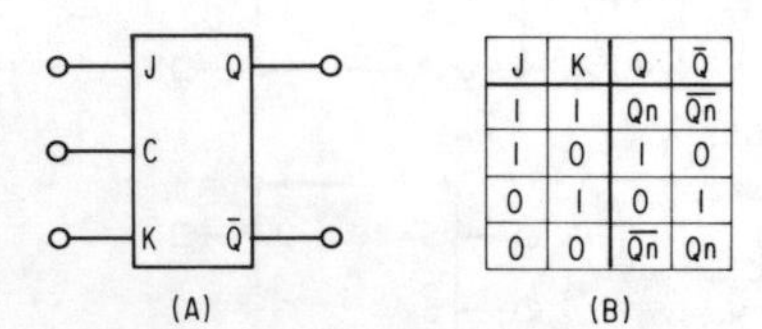

J	K	Q	$\bar{Q}$
1	1	Qn	$\overline{Qn}$
1	0	1	0
0	1	0	1
0	0	$\overline{Qn}$	Qn

Fig. 2-16. Typical JK flip-flop. (Courtesy Fairchild Semiconductor)

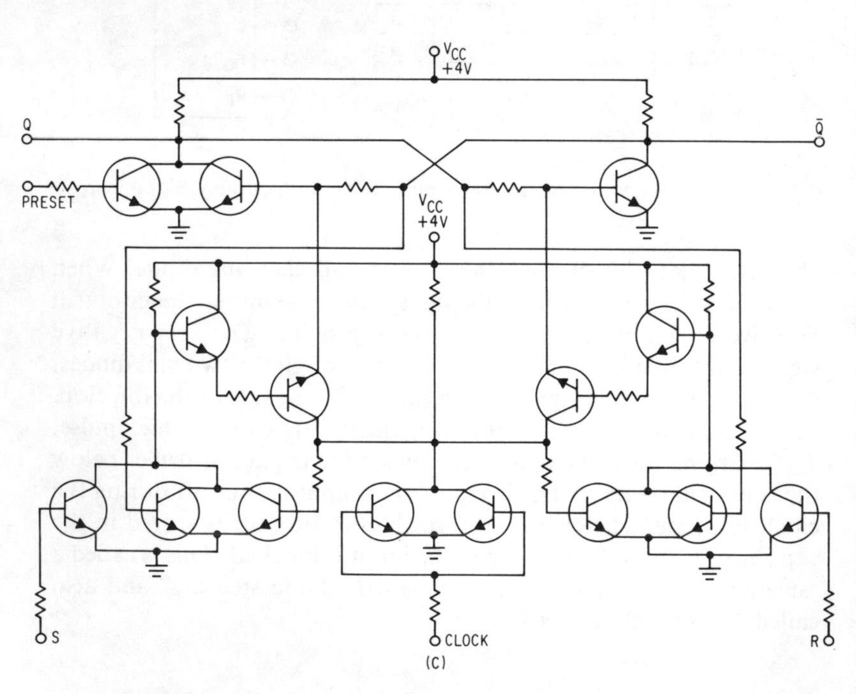

input, the set signal that isn't blocked is the one that changes the state of the flip-flop. This is illustrated in the truth table which shows that two 1s cause the flip-flop to assume the inverse of its previous state (Fig. 2-16B).

The three transistor gates at the lower right and lower left of the circuit are used to gate the set signal and the inverse output signal with the clock (Fig. 2-16C). The clock enters these gates through an inverter which holds off the input to the flip-flop while the clock is positive. The output of the set gates drives an emitter follower which is connected to the base of a transistor whose emitter is connected to the flip-flop. The connecting transistors serve as temporary storage for the input data. The storage is accomplished by charge storage in the connecting transistor. During the clock pulse up-time, the collector of the coupling transistor is driven negative and the base is driven positive. Current flowing in the forward-biased, collector-base

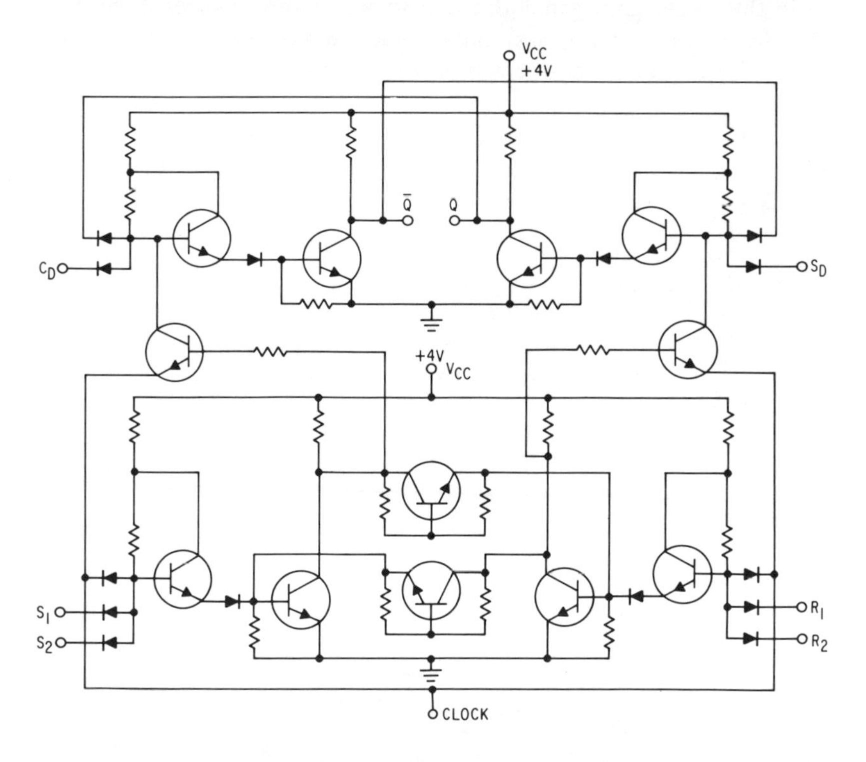

Fig. 2-17. Single-phase clocked RS or JK flip-flop. (Courtesy Fairchild Semiconductor)

junction causes a charge to be stored. When the clock goes negative, the clock inverter output goes positive, dumping the charge into the base of the flip-flop transistor. This circuit is sensitive to the rate of rise of the clock inverter output. However, the gain of the inverter reduces the sensitivity of the circuit to the clock signal fall-time and eliminates amplitude sensitivity.

Another level of sophistication is shown in the binary element in Fig. 2-17. Here, two complete flip-flops are used with conventional gating. The lower half of the circuit includes a flip-flop for temporary storage of the input. When the clock goes positive, the input is gated into this flip-flop via the diode gates. At the same time, a pair of clock inverters block the transfer of data from the input flip-flop to the storage flip-flop. When the clock goes negative, the input is shut off, and the input flip-flop transfers its contents to the storage flip-flop.

The basic types of popular integrated circuit flip-flops have been covered. There are many variations of these elements. Some simple flip-flops have gating included to form a half-shift register stage. Two of these would be approximately equivalent to one binary element. Other elements are connected in binary counter form.

3
Review of Semiconductors

Introduction

The bases of practically all of today's integrated circuits are two semiconductor materials, silicon and germanium. The fundamental electrical properties of both are similar. However, silicon components can operate at higher temperatures, and it is easy to "grow" an insulating film on top of silicon. Silicon, therefore, is in far greater use for today's IC fabrication. This chapter investigates those aspects of silicon semiconductor theory that relate directly to IC design and fabrication.

Crystals

Most solids, except those of cellular composition (bones, leaves, flesh, etc.) are crystalline in nature. Analyses have indicated that the atoms of these crystals are arranged in specific patterns, which depend on the size and number of atoms present and on the inter-atom electrical forces.

Figure 3-1 shows a pure silicon crystal. Each sphere represents a silicon atom with the four valence electrons normally found in the atom's outer shell, as shown in Fig. 3-2. In other words, the sphere contains the nucleus of the atom (fourteen protons/fourteen neutrons), ten tightly-bound electrons on the two inner rings, and four more-loosely-held electrons on the outer ring. (Germanium, with its thirty-two proton/neutron nucleus and twenty-eight electrons, also has four valence electrons available.) It is these four "loose" electrons that are shown in Fig. 3-2.

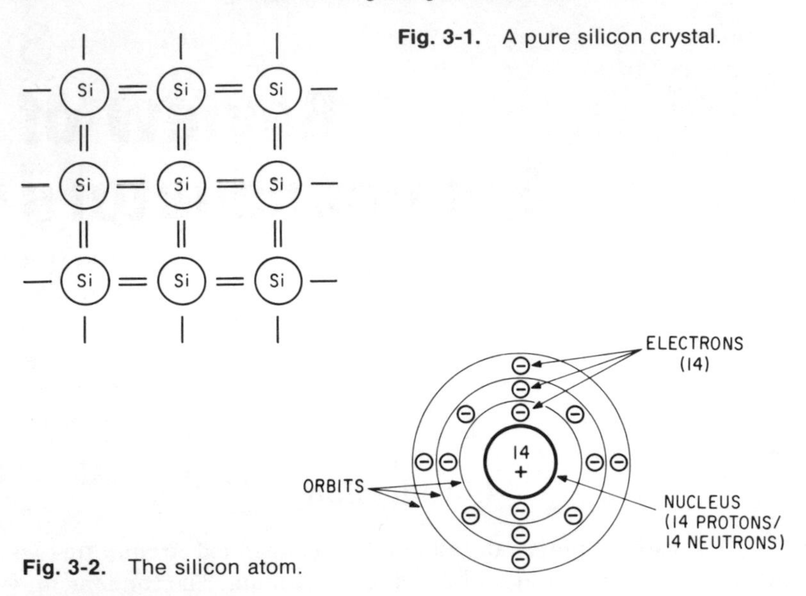

Fig. 3-1. A pure silicon crystal.

Fig. 3-2. The silicon atom.

Electrons rotate in relatively fixed orbits about the nucleus. In a crystal, the rotation of one valence (loose) electron of a given atom is coordinated with the rotation of one valence electron of an adjacent atom. This results in the formation of an electron-pair bond, also known as a valence bond. Crystalline materials, such as silicon and germanium, which contain these bonds, are poor conductors under normal conditions. Only if such materials are subjected to high heat or to electrical potentials will the bonds separate and permit partial electrical conduction.

It is possible for the atoms of other substances to join the silicon crystal structure. These atoms, whether found naturally, or added intentionally during IC fabrication are known as *impurities*. Two groups of impurities are capable of joining the silicon crystal. The first group, called *donors*, consists of materials such as arsenic, phosphorus, and antimony, all of which have *five* valence electrons. The second group, called *acceptors*, consists of materials such as aluminum, indium, and gallium, all with *three* valence electrons.

Donors, n-Type Silicon

Figure 3-3 shows a silicon crystal wherein one of the silicon atoms has been replaced by a donor impurity containing five valence electrons. The dark sphere represents the donor nucleus and all the tightly-bound electrons orbiting the nucleus; the valence electrons are not included in the sphere. Of the donor's five valence electrons, four form electron-pair bonds with electrons of neighboring silicon atoms. The donor's fifth valence electron cannot form a bond, as the silicon atom has only four valence electrons available. The result is that the donor's fifth electron becomes an excess electron. The donor nucleus has very little effect on this electron; at an ambient temperature of about 70° F enough heat energy is present to cause the excess electron to break away from the donor nucleus and wander through the crystal lattice, as shown in Fig. 3-3.

When the excess electron leaves (is donated by) the donor atom, the latter is left with a positive charge equivalent to the negative charge of one electron. The donor, therefore, is now an *ion.*

Silicon containing donor impurities is referred to as *n-type* silicon, the letter *n* referring to the negative charge of the excess electron.

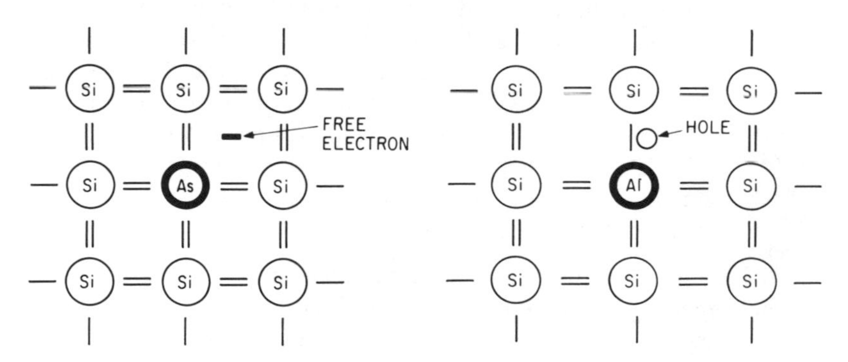

Fig. 3-3. A silicon crystal with arsenic added.

Fig. 3-4. A silicon crystal with aluminum added.

Acceptors, p-Type Silicon

Figure 3-4 shows a silicon crystal in which a silicon atom has been replaced by an acceptor impurity containing three valence electrons. The dark sphere represents the acceptor nucleus and all the tightly-bound orbiting electrons. The valence electrons are not included in the sphere. All the acceptor's three valence electrons form electron-pairs with the electrons of neighboring silicon atoms. Now we have silicon electrons with no place to go. Thus, an *electron-hole* arrangement exists. The position that would normally be filled by an electron is designated a *hole*.

For a good understanding of semiconductor theory, it is convenient to treat the hole as a specific particle. Holes in motion (described later) constitute an electrical current to the same extent that electrons in motion do. There are important differences, however, as follows:

1. The hole can exist only in a semiconductor material, as it depends for its existence on the electron-pair feature of crystal structures. Holes do not exist in conductors.
2. Holes are deflected by magnetic and electric fields in the same manner as are electrons. However, the hole has a positive charge, so that it moves in a direction opposite to that of an electron. In other words, the hole moves toward the negative pole of an electric field, whereas the electron moves toward the positive pole.
3. In electronics, the electron is considered indestructible. When a hole is filled by an electron from a nearby electron-pair, the hole is considered as having moved from one position to another. On the other hand,.when the hole is filled by a free, or excess electron, the hole no longer exists. The fact that silicon containing an equal number of donor and acceptor atoms has neither n- nor p-type characteristics bears out the latter statement.

If an electron from an adjacent electron-pair bond absorbs enough energy, it will break its bond. Figure 3-5 illustrates this. Figure 3-5A shows a p-type semiconductor just before the battery is switched into the circuit. With the battery connected, as in Fig. 3-5B, the electron from the adjacent covalent bond moves from its position in A and fills the hole, creating a new hole. When the hole moves to its new position, two important changes occur: (1) the acceptor atom has been ionized, i.e., it has acquired (accepted) an electron and is now a negatively charged ion; and (2) the silicon atom, which requires four valence electrons, is left with only three. Thus, lacking an electron, the atom has a net positive charge equal to the negative charge of one electron. Due to the electron-pair bond structure, this positive charge is neither scattered nor diffused; it is concentrated in the hole. Further, the positive hole moves within the crystal in the same manner that a free electron moves within the crystal.

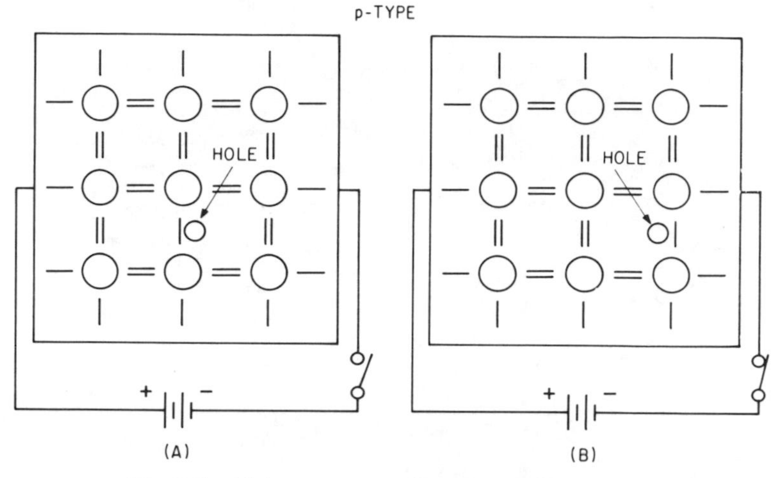

Fig. 3-5. Hole movement. (Courtesy RCA Corp.)

p-n Junctions

An unusual but very important phenomenon takes place at the surface where contact is made in the crystal between n-type and p-type silicon. The contact, known as a *p-n junction*, permits the use of semiconductors in circuits normally employing electron tubes.

Figure 3-6 shows separated sections of p-type and n-type silicon. The electron-pair bonds are not shown; only the holes, excess electrons, silicon spheres, and donor and acceptor ions are illustrated. The figure shows a large number of both acceptor and donor ions. Actually, however, the ratio of impurity to silicon atom is approximately 1 to 10 million.

Inside the silicon, the spheres and impurity ions vibrate within their lattice positions because of thermal energy. However, they do not leave their positions; hence, they do not constitute an electric current and may be considered to be fixed. On the other hand, the holes and excess electrons move randomly within the silicon. This movement is also due to thermal energy and is called *diffusion.*

Bonding p- and n-type materials together results in creating a depletion region as shown in Fig. 3-7. Normally, one would expect the holes in the p-type and the electrons in the n-type silicon to flow toward each other, combine, and eliminate all holes and excess electrons. Some do combine, with each combination eliminating a

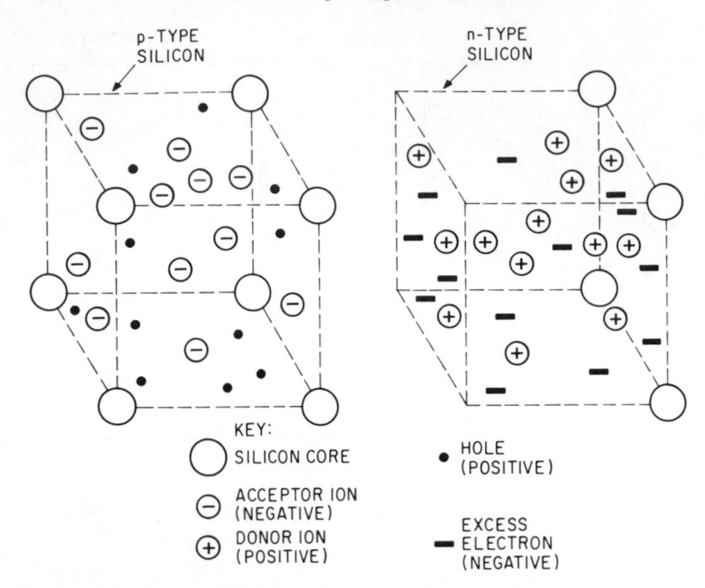

Fig. 3-6. A separated p-n junction.

hole and an *excess* electron. But this action takes place for a short time, and only in the junction's immediate vicinity. Negative acceptor ions in the p-region and positive donor ions in the n-region farther away from the junction are left uncompensated. Thus, additional holes that might diffuse into the n-region are repelled by the uncompensated positive charge of the donor ions. The same holds true for the electrons and uncompensated negative charges on the p-region acceptor ions. As a result, total electron-hole recombination cannot occur. The potential existing at the junction is called the *junction barrier*, the *potential gradient*, or the *potential barrier*.

Fig. 3-7. Bonding of p- and n-type materials creating the p-n junction.

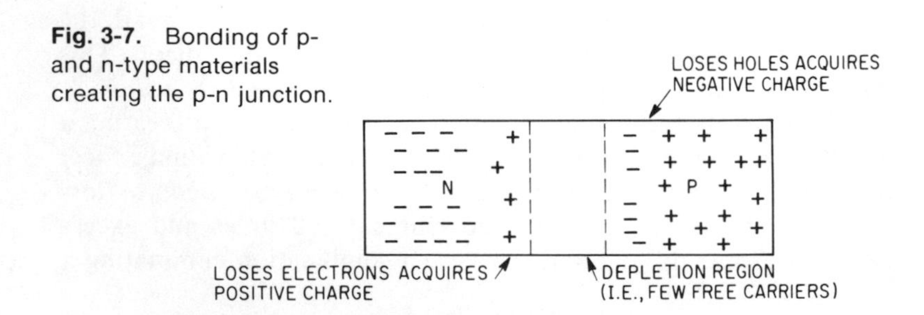

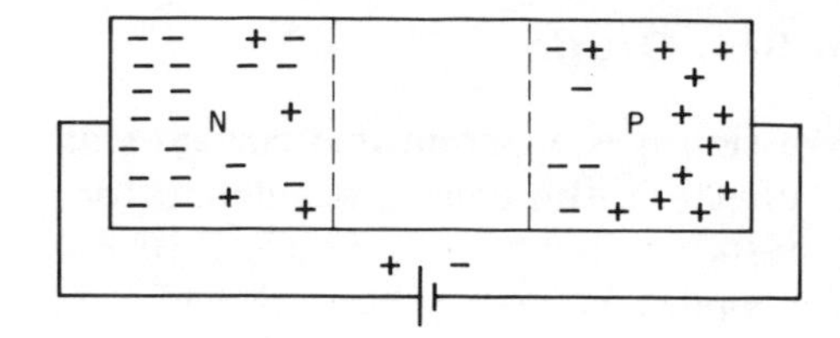

Fig. 3-8. The reverse bias connection of a p-n semiconductor creating a p-n junction.

Reverse Bias

Suppose a battery were connected across the p-n junction, as shown in Fig. 3-8. The holes would then drift away from the junction toward the negative terminal, and the electrons would move toward the positive terminal. This action would continue only until the potential barrier equalled the potential of the external battery. At this point, current flow (electron and hole movement) would stop, and the p-n junction would be biased in a *reverse* direction.

Forward Bias

If the external battery were now to be connected, as shown in Fig. 3-9, another phenomenon would occur. The holes would be repelled by the positive battery terminal and would move toward the p-n junction. Similarly, the electrons would be repelled by the negative terminal and also move toward the junction. With their newly acquired energy, some of these electrons and holes would combine. For each of these combinations, an electron would leave the battery's negative terminal, enter the n-type silicon, and move toward the junction. At the same time, the attractive force of the battery's positive terminal would rupture one of the electron-pair bonds in the p-type silicon. The electron set free would then enter the battery's positive terminal, leaving a hole. The hole, in turn, would drift and combine with an excess electron. This recombination process would continue as long as the battery remained connected, resulting in a constant flow of current. This is the condition known as *forward bias*.

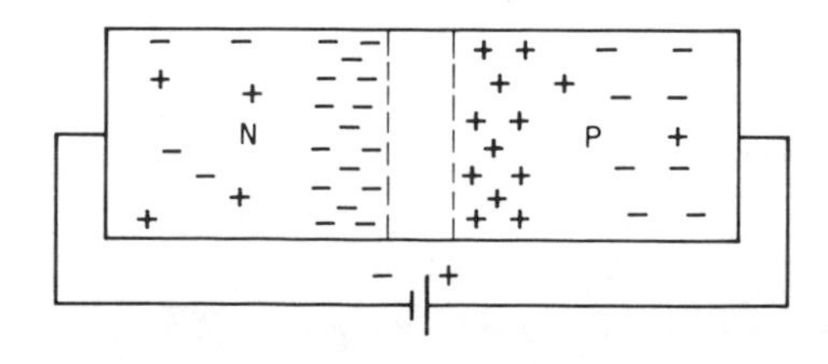

Fig. 3-9. The forward bias connection of a p-n semiconductor creating current flow.

Junction Diode

From the foregoing discussion it is apparent that current will flow more readily in one direction through a semiconductor, depending upon the polarity of the applied voltage. Figure 3-10 is a graph showing just how the applied voltage affects current in a junction diode (any single p-n junction is effectively a junction diode). In the forward direction the current is quite large, on the order of milliamperes (mA); whereas in the reverse direction, current is usually on the order of microamperes (μA). This small amount of reverse current which, from the previous discussion, would be expected to be zero, is due to the fact that neither the n- nor the p-type silicon is "pure." That is, there are always some acceptor ions and associated holes in the n-type silicon, and some donor ions and associated excess electrons in the p-type silicon. Since they are so few in number, the holes in the p-type silicon and the excess electrons in the n-type are called *minority carriers,* as opposed to the *majority carriers.*

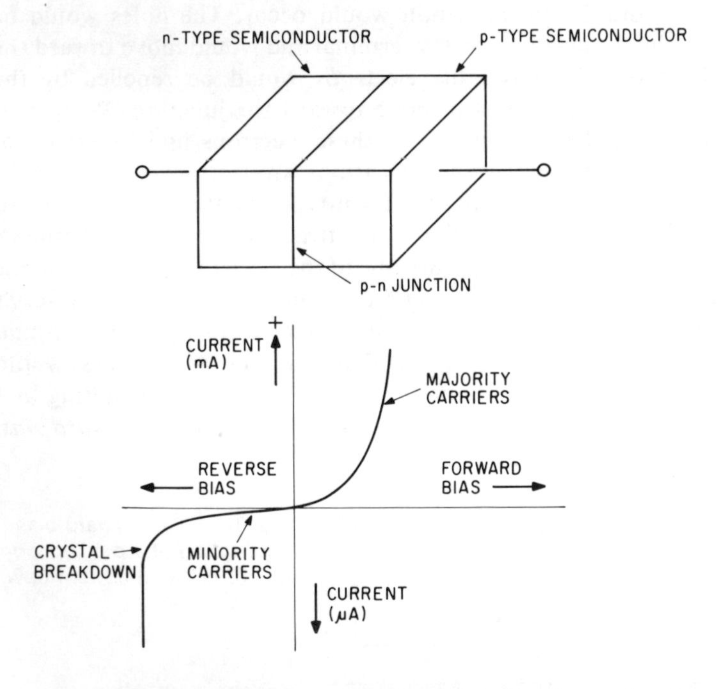

Fig. 3-10. The voltage and current relationship at a p-n junction.

The sudden negative increase in the reverse current, shown on the graph, occurs when the reverse bias is greater than the greatest possible potential barrier. This surge of high reverse current is not due to the minority carriers; it occurs because the crystal structure itself breaks down. This crystal breakdown is often referred to as the *zener breakdown* or *avalanche breakdown.*

Junction Transistor

Referring again to Fig. 3-10, note that a forward-biased p-n junction is equivalent to a low-resistance element, i.e., it will pass a high current for a given voltage. In turn, the reverse-biased p-n junction is the equivalent of a high-resistance element. From Ohm's Law for Power, $P = I^2R$, the power developed across a high resistance is greater than that across a low resistance. Thus, if a crystal were to contain *two* p-n junctions, a low-power signal could be injected into the forward-biased junction and produce a high-power signal that could be extracted from the reverse-biased p-n junction. In effect, a power gain would be obtained by *transferring* the signal from a small to a large *resistor.* Hence, the term *transistor,* the combination of *trans*fer and res*istor.*

p-n-p Transistor

Three silicon sections make up two p-n junctions. When combined as shown in Fig. 3-11 they form transistors, either the p-n-p type shown on the left or the n-p-n type shown on the right. Both are termed *bipolar*, as the performance of both depends on the interaction of two types of charge carriers, holes and electrons.

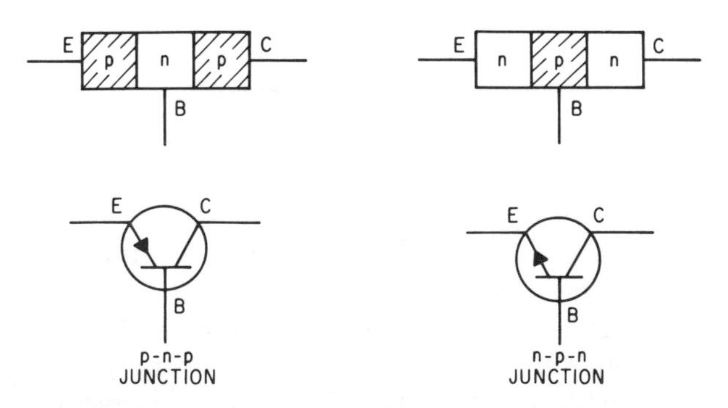

Fig. 3-11. P-n-p and n-p-n transistor configurations.

Figure 3-12 shows a p-n-p transistor with both forward and reverse biasing. The emitter is the forward-biased element, and the collector is the reverse-biased element.

Because of the simultaneous biasing, a great number of the emitter holes do not combine with the base electrons given off by the emitter-base battery. Many of these holes diffuse through the base into the base-collector region where they combine with electrons from the base-collector battery. Since maximum transistor power gain (greatest power output for any specific input) depends on maximum diffusion through the base into the collector, practical transistors are made with the base material very narrow as compared to the emitter-collector materials.

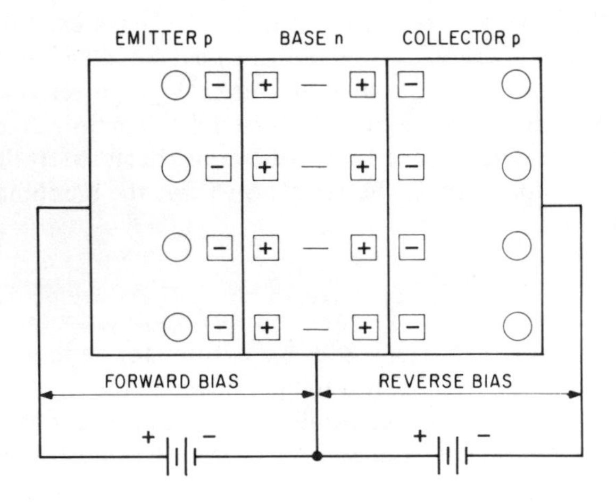

Fig. 3-12. A p-n-p junction transistor circuit.

n-p-n Transistor

With two major differences, the n-p-n transistor operates in a manner similar to that of a p-n-p. These differences are:

1. Whereas the current carrier in the p-n-p transistor is the hole, in the n-p-n type it is the electron.
2. The bias voltages (A and B in Fig. 3-13) in an n-p-n transistor are the reverse of those in the p-n-p. This is obviously necessary, since different materials are now being used for the emitter, base, and collector, respectively.

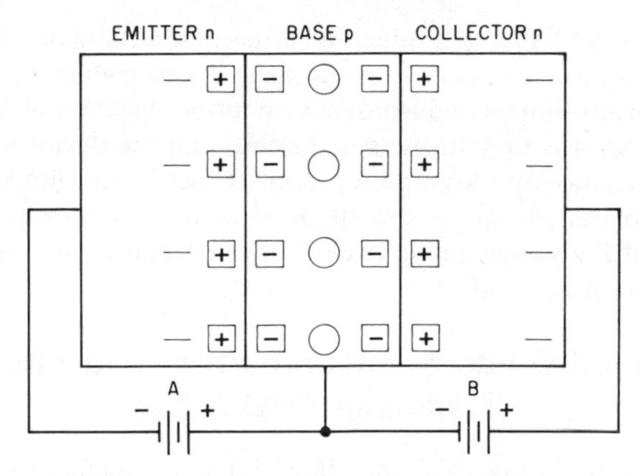

Fig. 3-13. An n-p-n junction transistor circuit.

Junction Field-Effect Transistor (JFET)

Although practically obsolete, as it has been replaced by the metal-oxide semiconductor field-effect transistor (MOSFET), a brief discussion of the JFET will round out the material on p-n junction devices.

The operation of field-effect devices can be explained in terms of a charge-control concept. Referring to Fig. 3-14, the metal control electrode, the gate, acts as a charge-storage or control element. A charge on the gate induces an equal but opposite charge in the semiconductor layer, or *channel,* located beneath the gate. The charge induced in the channel can then be used to control the conduction between two ohmic contacts, the source and drain, made to opposite ends of the channel.

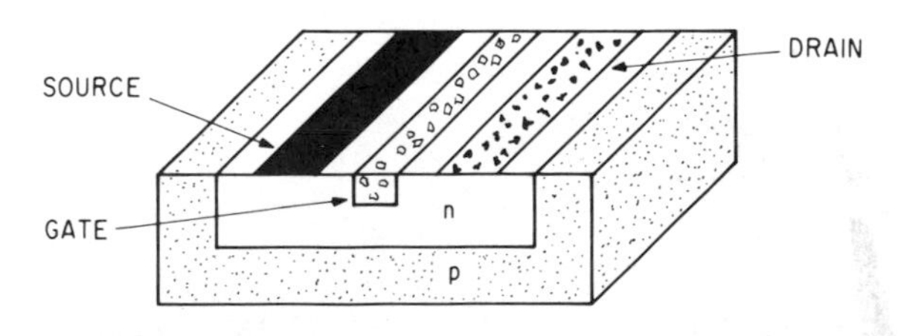

Fig. 3-14. Structure of a p-n junction field-effect transistor. (Courtesy RCA Corp.)

In the JFET, a p-n junction is used for the gate. When the junction is reverse-biased, it functions as a charge-control electrode. Under steady-state conditions, i.e., continuous reverse bias, only leakage currents flow in the gate circuit, and the device has a high input resistance and high power gain. If the device should become forward-biased, however, the input resistance then drops sharply. The result is an appreciable rise in input current with a subsequent decrease in power gain.

Metal-Oxide Semiconductor Field-Effect Transistor (MOSFET)

As shown in Fig. 3-15, the MOSFET uses a metal gate electrode that is separated from the semiconductor material by an insulator. Like the p-n junction, this insulated-gate electrode can deplete the source-to-drain channel of active carriers when suitable bias voltages are applied. However, the insulated gate electrode can also increase the conductivity of the channel without increasing steady-state current or reducing power gain.

Two basic types of MOSFET exist: *depletion* and *enhancement*. In the depletion type, charge carriers are present in the channel, and the channel is conductive when no bias is applied to the gate. A reverse gate voltage is one which depletes this charge and thereby reduces channel conductivity. A forward bias draws more charge carriers into the channel, thus increasing channel conductivity.

In the enhancement type, the gate must be forward-biased to produce active charge carriers and permit channel conduction. No useful channel conductivity exists at either zero or reverse gate bias.

As MOSFETs can be made to utilize electron conduction (n-channel) or hole conduction (p-channel), four types of MOSFET are

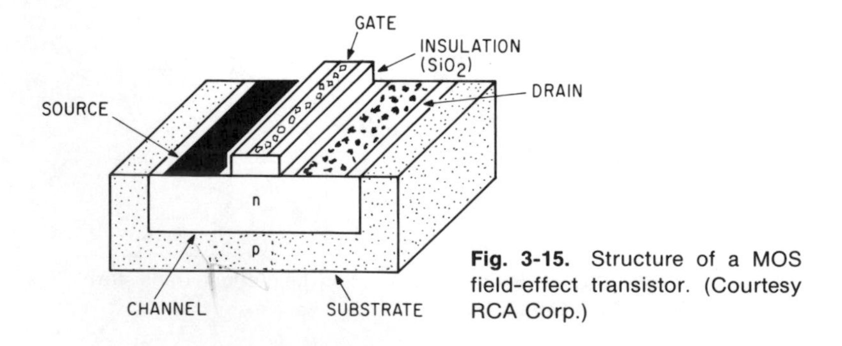

Fig. 3-15. Structure of a MOS field-effect transistor. (Courtesy RCA Corp.)

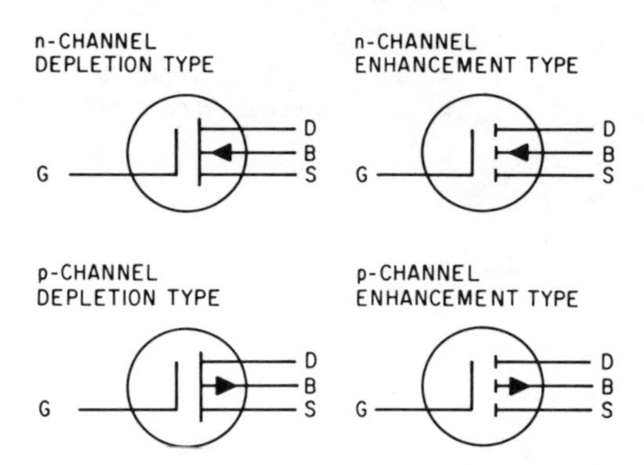

Fig. 3-16. Schematic symbols for MOS transistors. (G = gate; D = drain; S = source; B = active substrate; B is not generally indicated on schematics.) (Courtesy RCA Corp.)

possible. The symbols shown in Fig. 3-16 show whether the transistor is n-channel, p-channel, depletion, or enhancement type. The direction of the B arrowhead in the symbols identifies the channel type: arrowhead inward connotes an n-channel type; arrowhead outward connotes a p-channel type. Enhancement or depletion is identified by either a dotted channel line or a solid channel line, respectively.

Not mentioned before is the fact that both the JFET and the MOSFET are *unipolar* devices, as opposed to the bipolar n-p-n and p-n-p transistors. The term unipolar means that operation is basically a function of only *one* type of charge carrier: holes in p-channel devices, and electrons in n-channel devices.

If we were to reverse the type of conductivity in a MOSFET, the resulting device would be "complementary" in characteristics to the

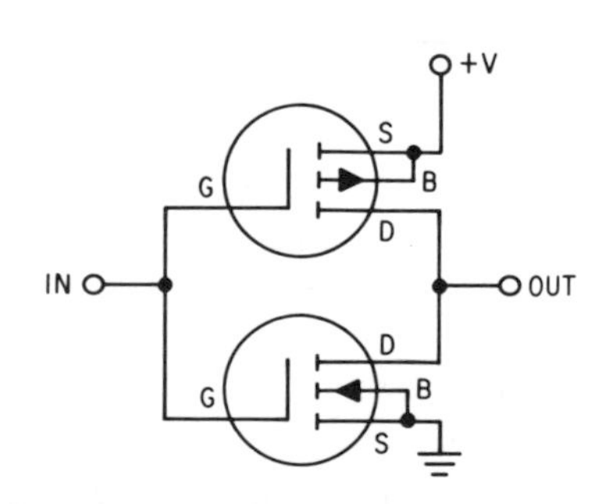

Fig. 3-17. Complementary-symmetry inverter circuit using MOS transistors.

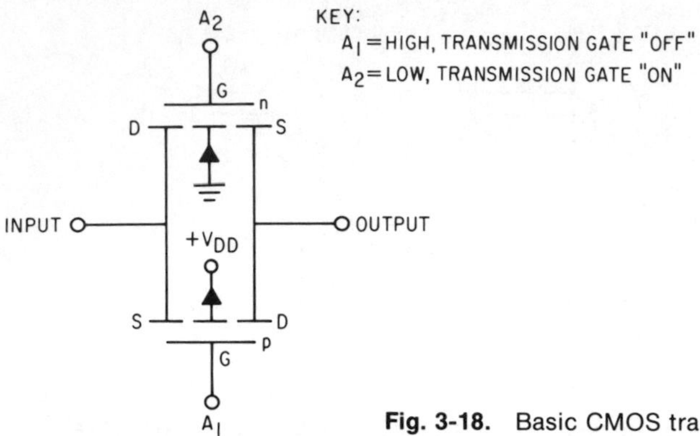

Fig. 3-18. Basic CMOS transmission gate.

original device. That is, n-channel MOSFETs are related to p-channel MOSFETs in the same manner as p-n-p transistors are related to n-p-n transistors. So, if we were to connect "complementary" MOSFETs in a circuit like that of Fig. 3-17, we would have a complementary circuit. Thus we have one of the basics of complementary MOS (CMOS), a simple complementary inverter. The other CMOS basics are the transmission, NOR, and NAND

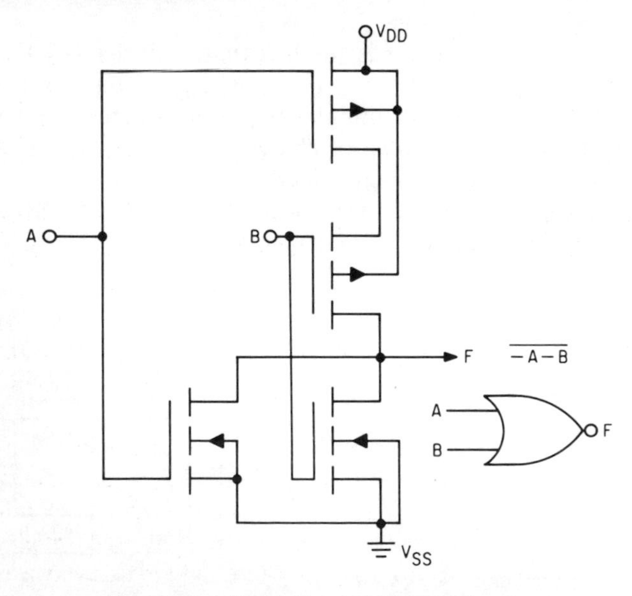

Fig. 3-19. A two-input CMOS NOR gate.

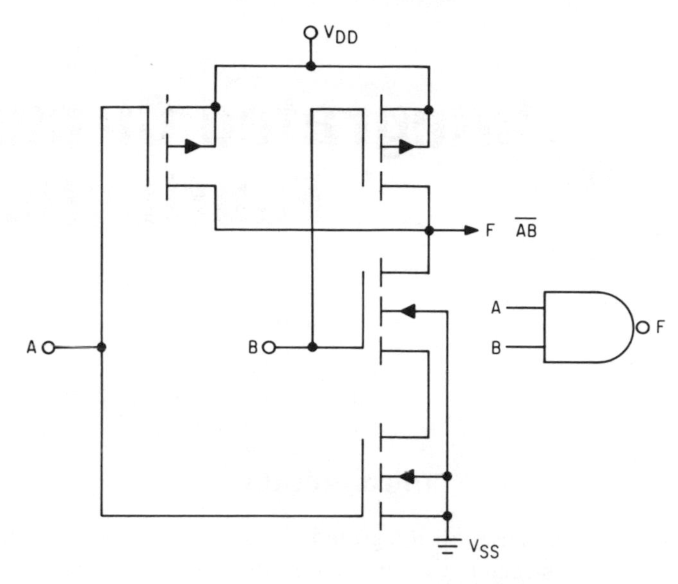

Fig. 3-20. A two-input CMOS NAND gate.

gates. The transmission gate, basically, is a single-pole, single-throw switch that has a very high off-to-on resistance ratio—typically on the order of 10^9. The circuit is formed by connecting two complementary MOS transistors in parallel, as shown in Fig. 3-18. When the control signals are applied as indicated, the circuit functions as a bidirectional switch and provides either low *on* or high *off* impedance to signals between ground and $+V_{DD}$.

A NOR gate is formed by connection of two or more paralleled n-channel transistors in series with two or more series-connected p-channel transistors. Figure 3-19 shows the circuit diagram for a two-input NOR gate. A negative output is obtained when either input A or input B is positive. For each condition, the n-channel transistors are turned on and the p-channel transistors are turned off, so that the output is connected to V_{SS} (ground).

A NAND gate consists of two or more series-connected n-channel transistors in series with two or more parallel-connected p-channel transistors. Figure 3-20 shows the circuit diagram of a two-input NAND gate. The output of this circuit is low (V_{SS} or ground) only if both inputs are high (V_{DD}), because only in this way are both n-channel transistors turned on to connect the output to V_{SS}.

4
Integrated Circuit Fabrication

Introduction

In a broad sense, all integrated circuits (ICs) have at least one identifying characteristic in common: Active and passive components are combined or interconnected on a common substrate to obtain a complete circuit function. This approach contrasts with assembling discrete circuits, no matter how miniaturized they may be. ICs, perhaps, received their greatest impetus from the highly sophisticated semiconductor technology available today. This technology has resulted in high-volume production of monolithic and MOS circuits. It must be realized, however, that the term "integrated circuits" has broad applicability.

The exact structure and interconnect method used in fabricating an IC can be useful in classifying the IC. Several different basic technologies can be employed, as shown in Fig. 4-1. An IC can be fabricated by any one of the processes indicated, or by any combination thereof. The actual choice depends upon many factors, such as the functional end use, the type of application (commercial, industrial, or military), production volume, and cost.

Both thick- and thin-film circuits are usually constructed on *insulating* substrates of ceramic or glass. Monolithic bipolar and MOS (unipolar) circuits are constructed on *semiconducting* substrates. Dielectric isolation and beam-lead devices employ the desirable features of both film and semiconductor circuits, without the shortcomings of either.

A multi-chip circuit is made from two or more separate semiconductor substrates interconnected in the same package. A hybrid IC, on the other hand, is made from any combination of

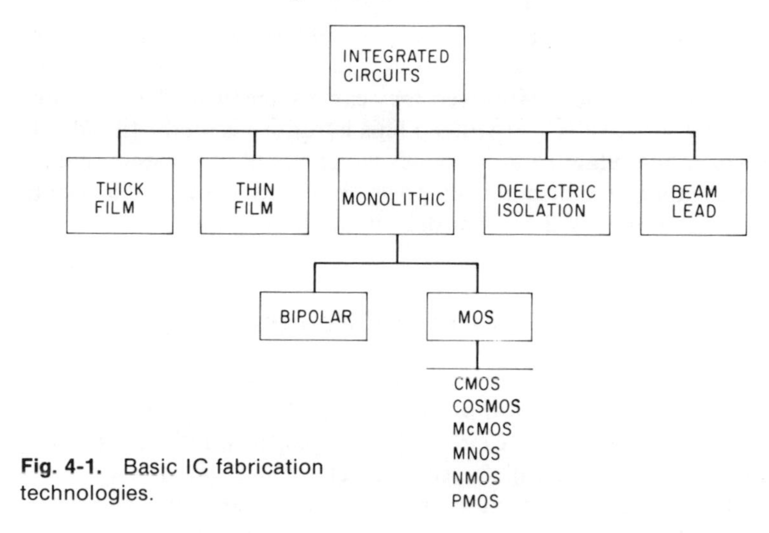

Fig. 4-1. Basic IC fabrication technologies.

individual technologies, and may on occasion contain conventional discrete components.

Thick-Film Circuits

(The reader should refer to Chapter 1 for a brief comparison of thick- and thin-film circuits.) The actual fabrication steps in a typical thick-film process are as follows.

First, a schematic diagram is evaluated with respect to its suitability to the thick-film process. An oversize drawing is made of the layout, and is then reduced photographically. Subsequently, metal screens are produced and fitted to printers. The paste or "ink" that is to be printed onto the substrate consists of a mixture (slurry) of conductors (or nonconductors, if the film is to be dielectric) and pulverized glass or ceramic suspended in a suitable liquid carrier. The ink is then squeegeed through predetermined openings in the screen onto the substrate. The process is then repeated as necessary for conductive, resistive, and possibly dielectric paths. Finally, the substrate is dried and then heated in a furnace to approximately 500°-1200°C. This process forms a molecular bond between the glass or ceramic constituents of the slurry and the substrate.

Resistors made this way have values that may vary as much as 10 to 50 percent, but can be adjusted by abrasive trimming to tolerances as low as 0.3 percent.

Thick-film circuits are usually combined with semiconductor circuits to give complete hybrids, such as high-voltage divider networks for TV sets. However, they can also be formed into ICs and as purely-resistive IC networks. (This latter use is on the rise, due to improved manufacturing processes which result in resistors that are both cheaper than discrete composition resistors, and just as stable, i.e., a maximum of 2 percent drift in value with age.)

Thin-Film Circuits

Because of the film's thinness, raw ceramic materials are not smooth enough for their application, and must, therefore, be glazed. As a result, substrates are either glass or glazed Alumina, depending upon the thermal and dielectric characteristics desired.

Early thin-film technology produced thin-film transistors (TFT). However, present technology has found that only passive devices are commercially feasible, although a new IBM process may find thin films being used to replace LEDs (light-emitting diodes) in

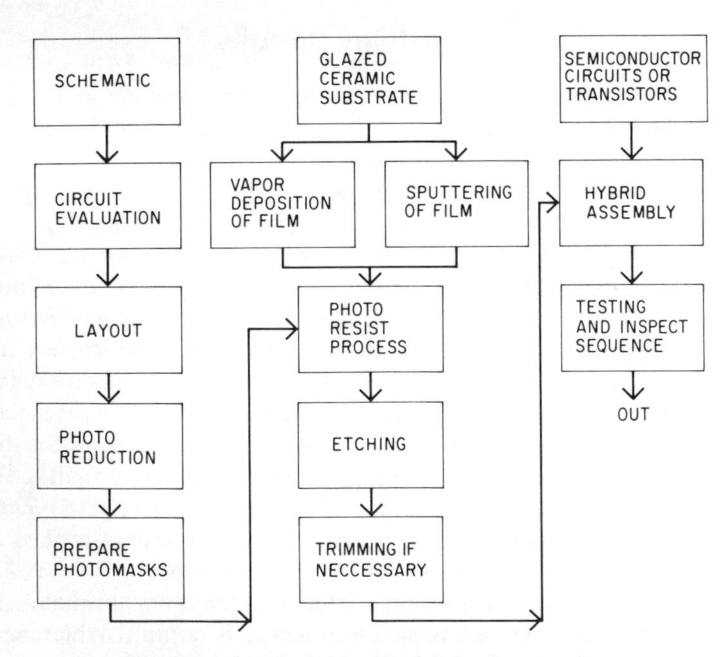

Fig. 4-2. A typical thin-film fabrication process.

display devices. A typical thin-film fabrication process is shown in Fig. 4-2.

As the film is deposited on the entire substrate, possibly in several layers, the unwanted portions must be removed by subtractive etching. A photoresist is applied to define the areas to be etched. The substrate is spun at high speed on a turntable during the photoresist application to assure uniform coating. The photoresists themselves are solutions of organic compounds (lacquers) that are sensitive to ultra-violet light.

The resist is permitted to dry, and is then exposed through a suitable photomask to ultra-violet light. The unwanted resist is then removed with an appropriate developer. The unwanted metal is then removed by means of a suitable etchant, one which will attack only the film and not the substrate.

Thin-film resistors can be very accurately defined, with typical possible line widths of 0.0013 cm ±5 percent absolute accuracy without trimming. In addition, thin films can be deposited on top of the silicon dioxide surface of planar monolithic circuits, and be compatible with the latter.

Common sheet resistivities of nichrome, chromium, and tantalum films range from 500,000 ohms per square, with temperature coefficients of ±10 to 200 ppm/°C. Maximum dc operating potential for thin-film resistors is approximately 100V.

Monolithic Integrated Circuits

Introduction

The manufacture of silicon monolithic ICs involves a series of highly-critical operations in which the silicon is subjected to carefully-controlled, high-temperature chemical processes. The heart of the technology is *masked surface diffusion*, involving temperatures of about 1200°C. Monolithic ICs are "batch" fabricated, with a pure silicon "wafer" comprising a minimum batch. Individual silicon wafers about 10 mils thick are cut from an ingot of pure silicon crystal, then polished to a mirror finish by acidic etching.

The first major step in processing the polished wafer is to form a silicon dioxide layer on the wafer surface, as shown in Fig. 4-3. This is done by passing oxygen or steam over the heated (1000°–1300°C) silicon wafer until the desired wafer thickness is obtained. The oxide layer plays three roles in IC production: (1) it forms a mechanical

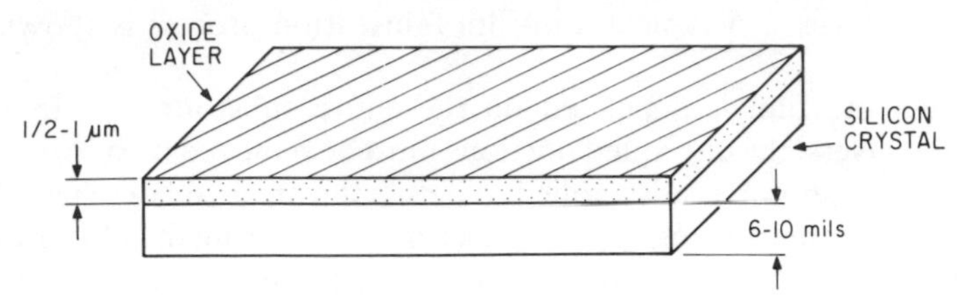

Fig. 4-3. Typical dioxide layering.

mask through which selective diffusion of impurities (dopants) occurs; (2) it acts as a barrier to dopants during the semiconductor-junction forming processes, i.e., it "passivates" the semiconductor surface; and (3) it provides an insulating substrate for the interconnection metals.

Following oxidation, the oxide layer is coated with a photoresist that has the ability to harden (polymerize) when hit by light. Next, a photomask, say a glass plate containing a pattern of opaque black spots, is placed against the photoresist, and the entire system is exposed to ultraviolet light.

The illuminated areas of the photoresist tend to polymerize, while the areas under the opaque black spots remain soft, for easy removal during the next step. The wafer is next subjected to a chemical etchant which dissolves the oxide in the windows without attacking the wafer itself; the result is the desired windows. The remaining photoresist is now dissolved completely via another chemical process, to prevent its contaminating the processed wafer during the remainder of the fabrication. The "cleaned" wafer, with the windows in its oxide coating, is now ready for chemical doping procedures in diffusion furnaces, in order to obtain regions with n- or p-type characteristics in the areas beneath the windows.

Chemical diffusion processes are used to introduce the semiconductor-junction-forming dopants into the bulk silicon.

Formation of Linear Circuit Components

Most of today's ICs are produced by so-called planar-epitaxial processes, i.e., processes wherein all the p-n junctions are buried within an epitaxial layer. (This is not the "buried" layer, which is discussed later.) This process has an advantage over the simple diffusion described previously, even though diffusion is still employed. In simple diffusion, some of the dopant gets into the base silicon; by utilizing an epitaxial layer, this possibility is eliminated.

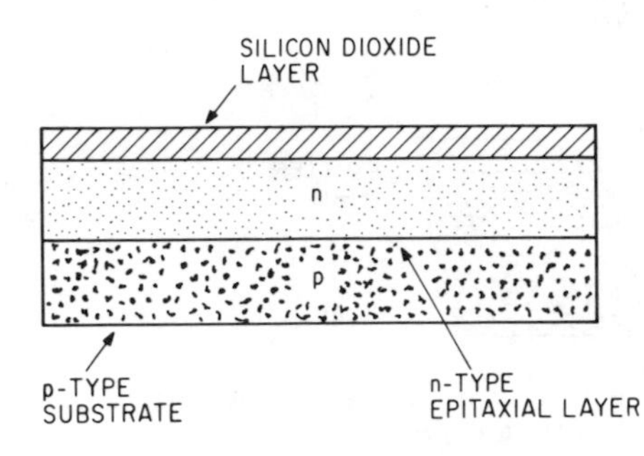

Fig. 4-4. Basic monolithic IC wafer. (Courtesy RCA Corp.)

The substrate for a monolithic IC consists of a uniform single crystal of n- or p-type silicon. An oppositely-doped (p- or n-type, respectively) epitaxial layer is "grown" (deposited) onto the substrate, and a very thin film of silicon dioxide (SiO_2) insulation is then formed, as already described, on the layer rather than directly onto the substrate. Figure 4-4 shows the resultant structure, assuming a p-type substrate. The first set of windows is then produced, as before. Next, p-type material is diffused vertically into the n-type epitaxial layer to form insulated n-type nodes in the regions in which the desired circuit components are to be located, as shown in Fig. 4-5. The diode junctions formed by the n-type nodes and p-type material provide electrical isolation between the nodes.

Low-resistance n+-type (the + sign indicates a much heavier concentration than normal) pockets, as shown in Fig. 4-5, are diffused into the p-type substrate immediately below the isolated n-type nodes. These pockets, also known as "buried layers," reduce the resistance of the n-type nodes so that they can be used as the collectors of IC transistors. To form such a transistor, additional p-type and n+-type regions are diffused within an isolated n-type node, as shown in Fig. 4-6. The additional n+-type material forms the

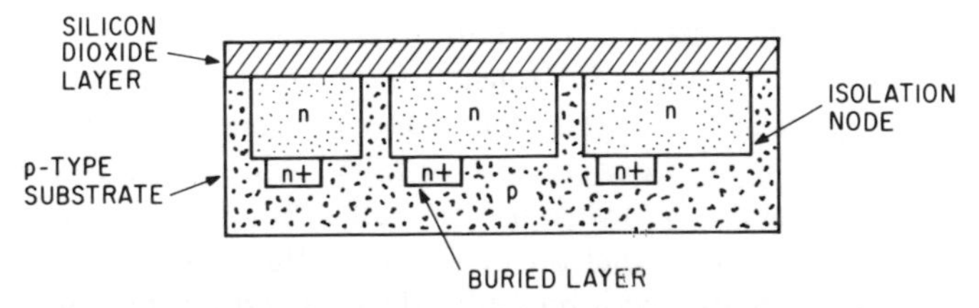

Fig. 4-5. Diffusing p-type material into n-type epitaxial layer to provide isolated n-type nodes. (Courtesy RCA Corp.)

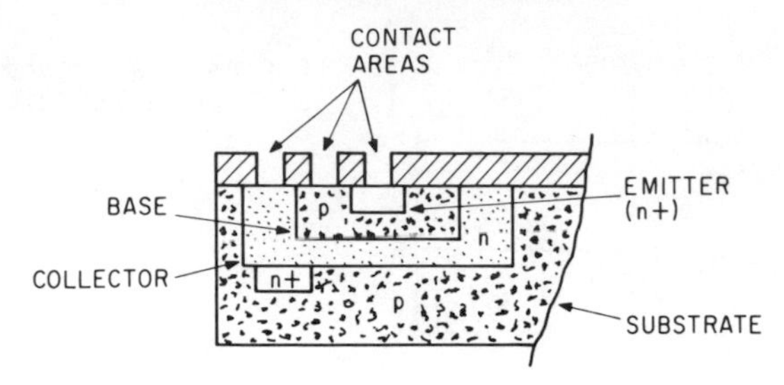

Fig. 4-6. Diffusing additional p- and n-type regions to form monolithic transistors. (Courtesy RCA Corp.)

emitter, while the additional p-type forms the base; the original n-type material, as already stated, forms the collector. The transistor just formed is a bipolar one.

A technique that reduces the size of IC transistors by 70 percent over the foregoing technology, and by 40 percent over the company's Isoplanar technology, was recently developed by Fairchild Semiconductor. Called Isoplanar II, the process is based on a walled-emitter structure claimed to reduce the transistor collector-base junction area by 50 percent over conventional designs. The Isoplanar process employs a silicon dioxide wall to reduce the isolation area required between adjacent transistors. Isoplanar II reduces the size of the transistor itself by terminating the ends of the base-emitter junction at the same wall that provides the isolation between transistors. Not only is packing density increased, but there is a significant reduction in junction capacity and power consumption, with an attendant increase in speed. Transistors made with this process exhibit reduced collector and isolation capacitance, but their IC breakdown voltage, beta, and open-base voltage equal those of conventional transistors made with the same diffusion levels.

Monolithic diodes may be formed by n-type diffusion into the silicon substrate. Such junctions, however, are essentially the same as the emitter-base junctions of monolithic transistors. When it is desirable to match diodes to transistors on the same silicon chip, a monolithic transistor is usually connected to operate as a diode. This type connection provides a low-voltage diode with low series resistance for general-purpose use as well as for biasing purposes.

The emitter-base junction of a monolithic transistor may be biased in the reverse direction to provide a zener diode (at about 7V), as shown in Fig. 4-7A. When lower dynamic impedance is desired, a

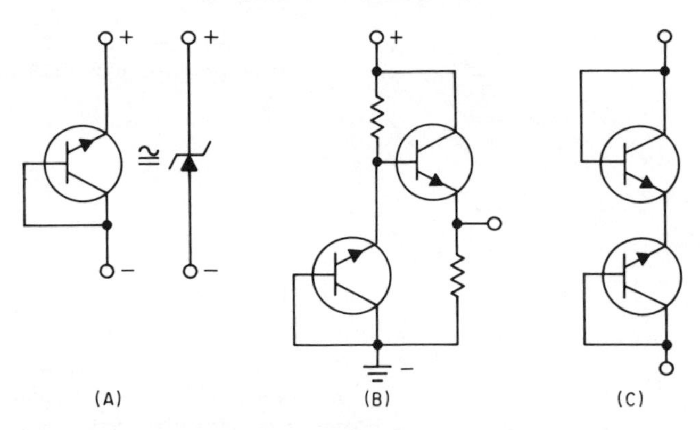

Fig. 4-7. Integrated zener diodes. (A) Using an emitter-base junction to provide a zener diode. (B) Using an emitter follower to provide lower dynamic impedance. (C) Temperature-compensated zener diode. (Courtesy RCA Corp.)

transistor may be used as an emitter follower for the diode, as shown in Fig. 4-7B. This configuration provides an equivalent zener voltage of approximately 6.3V. Figure 4-7C shows an arrangement for a higher-impedance zener diode that is almost temperature compensated; this arrangement provides a zener voltage of approximately 7.7V.

In monolithic ICs, only a few different resistivities can be provided. As a result, incremental values of resistance are determined primarily by the geometry of the resistors. The p-type resistor shown in Fig. 4-8 is the type most commonly used in monolithic ICs. This type of resistor is formed by a p-type base diffusion. The range of practical values for p-type diffused resistors is approximately 100–

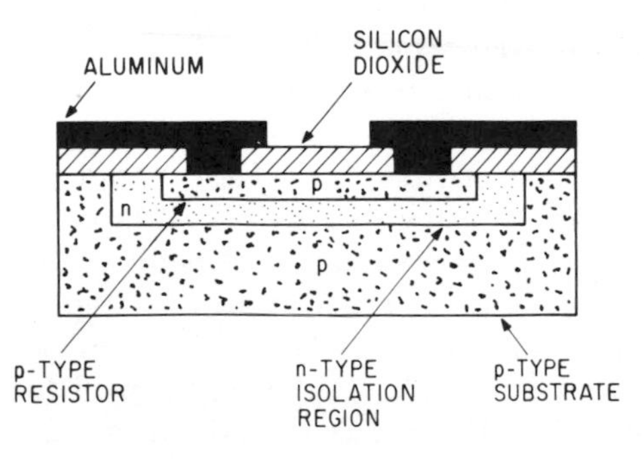

Fig. 4-8. Cross-section of a p-type diffused resistor. (Courtesy RCA Corp.)

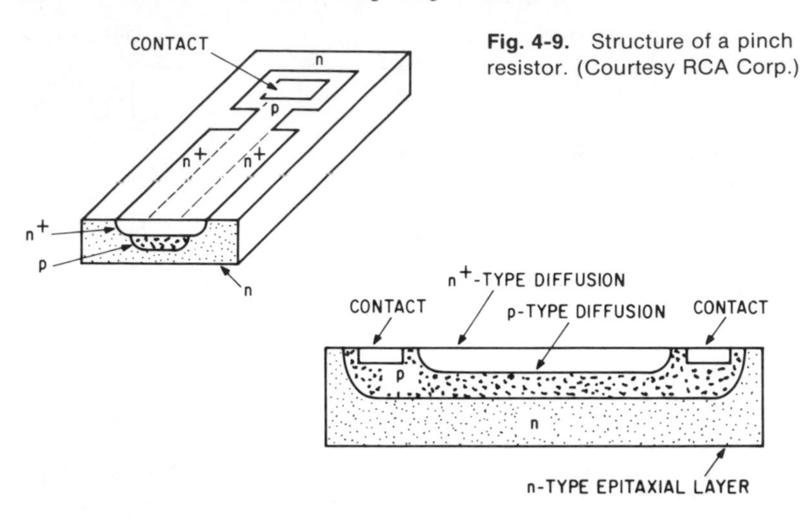

Fig. 4-9. Structure of a pinch resistor. (Courtesy RCA Corp.)

25,000 ohms. Higher values for monolithic resistors can be obtained by addition of an n+-type diffusion on the normal p-type diffused resistor, as shown in Fig. 4-9. The addition of the n+-type layer, in effect, reduces the cross-sectional area of the resistor and elminates the higher-conductivity path near the surface. Resistors of this type, termed "pinch" resistors, range from approximately 10,000 to 500,000 ohms.

The epitaxial layer of a monolithic IC may also be used to form high-value resistors. The length and width of this type of resistor are determined by the isolation diffusion. The practical range of epitaxial resistors is approximately the same as for pinch resistors. However, they require more chip area for a particular resistance than do the pinch types.

Monolithic capacitors can also be formed, and may be either diffused-junction types or metal-oxide semiconductor (MOS) types. (MOS capacitors are discussed further in the next section.) Any reverse-biased semiconductor junction has a depletion region which acts as a dielectric between two conductive surfaces. The capacitances of such junctions, however, vary with the reverse-bias voltage and with the physical size and doping concentration of the junction. For a reverse bias of 5V and the impurity concentration of a typical monolithic transistor, the capacitance per area is typically 0.07 pF per square mil for a collector-substrate junction, 0.1 pF per square mil for a base-collector junction, and 0.39 pF per square mil for an emitter-base junction.

Until recently, the fabrication of monolithic inductors was not considered feasible for any significant values of inductance. Among other factors, circuit losses are intolerably high. Instead, monolithic amplifier circuits are generally used in active networks to eliminate the need for inductors, which are large, heavy, and expensive. For example, IC operational amplifiers are being used in circuits that obey the laws of inductance, but require no large coils or metal cores.

This may soon change, however, for ceramic chip capacitor technology has been adapted by the San Fernando Electric Manufacturing Co. in California to produce a reliable chip inductor. Called Magna-Chip, the device is produced by screening u-shaped elements on individual layers to form a coil embedded in ferrite ceramic. This results in maximum coupling of the magnetic fields caused by current flow in the coil, thereby providing large values of inductance in a small volume. Because the winding is compact, its electrical resistance is low and its current-carrying capacity is high.

The chip inductors are comparable in size and reliability to chip capacitors. Inductance ranges from 0.2 to 5 μH, with a minimum Q of 20. Self-resonant frequency is greater than 50 MHz.

Formation of MOS Circuit Components

The metal-oxide semiconductor field-effect transistor (MOSFET), unlike the bipolar transistor, requires only one diffusion step in its simplest form, as shown in Fig. 4-10. Here, only two high-temperature operations are required, whereas ten are necessary for bipolars. The transistor shown is a pMOS type. By substituting p for n and n for p in the figure, the device becomes an nMOS type. (With the promise of semiconductor memories that will finally be cheaper

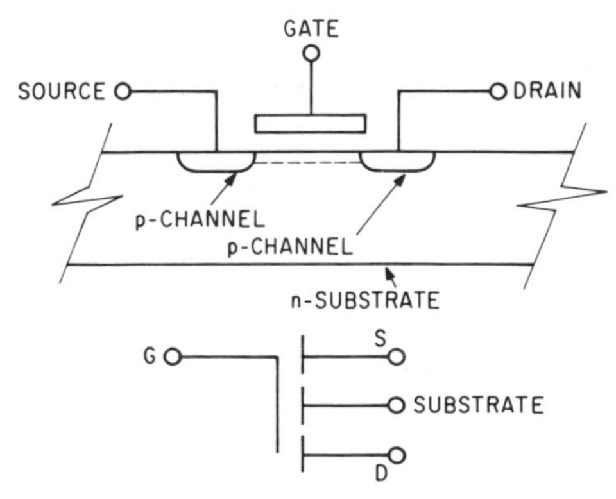

Fig. 4-10. Simple p-channel enhancement-mode MOSFET.

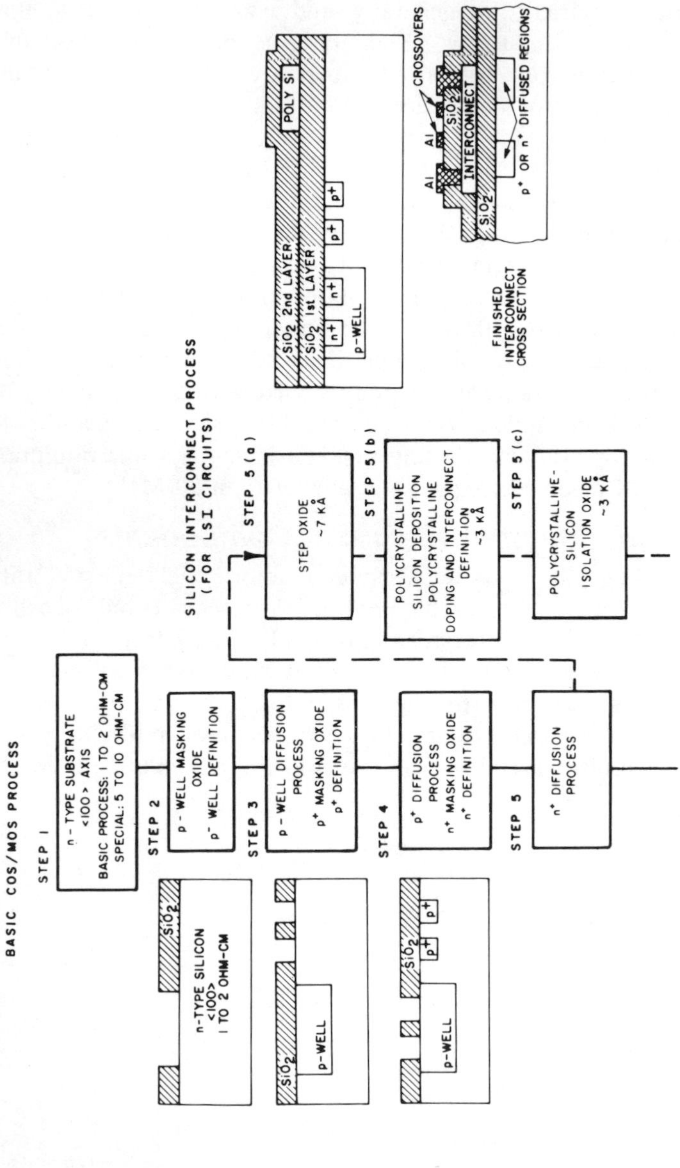

BASIC COS/MOS PROCESS

STEP 1

n - TYPE SUBSTRATE
<100> AXIS
BASIC PROCESS: 1 TO 2 OHM-CM
SPECIAL: 5 TO 10 OHM-CM

SiO_2
n-TYPE SILICON
<100>
1 TO 2 OHM-CM

STEP 2

p - WELL MASKING OXIDE
p^- WELL DEFINITION

STEP 3

p - WELL DIFFUSION PROCESS
p^+ MASKING OXIDE
p^+ DEFINITION

SiO_2
p-WELL

STEP 4

p^+ DIFFUSION PROCESS
n^+ MASKING OXIDE
n^+ DEFINITION

SiO_2
p-WELL
p+ p+

STEP 5

n^+ DIFFUSION PROCESS

SILICON INTERCONNECT PROCESS
(FOR LSI CIRCUITS)

STEP 5 (a)

STEP OXIDE
~7 KÅ

STEP 5 (b)

POLYCRYSTALLINE SILICON DEPOSITION
POLYCRYSTALLINE DOPING AND INTERCONNECT DEFINITION
~3 KÅ

STEP 5 (c)

POLYCRYSTALLINE-SILICON
ISOLATION OXIDE
~3 KÅ

SiO_2 2nd LAYER
POLY Si
SiO_2 1st LAYER
n+ n+ p+ p+
p-WELL

Al Al CROSSOVERS
SiO_2
INTERCONNECT
SiO_2
p^+ OR n^+ DIFFUSED REGIONS

FINISHED INTERCONNECT CROSS SECTION

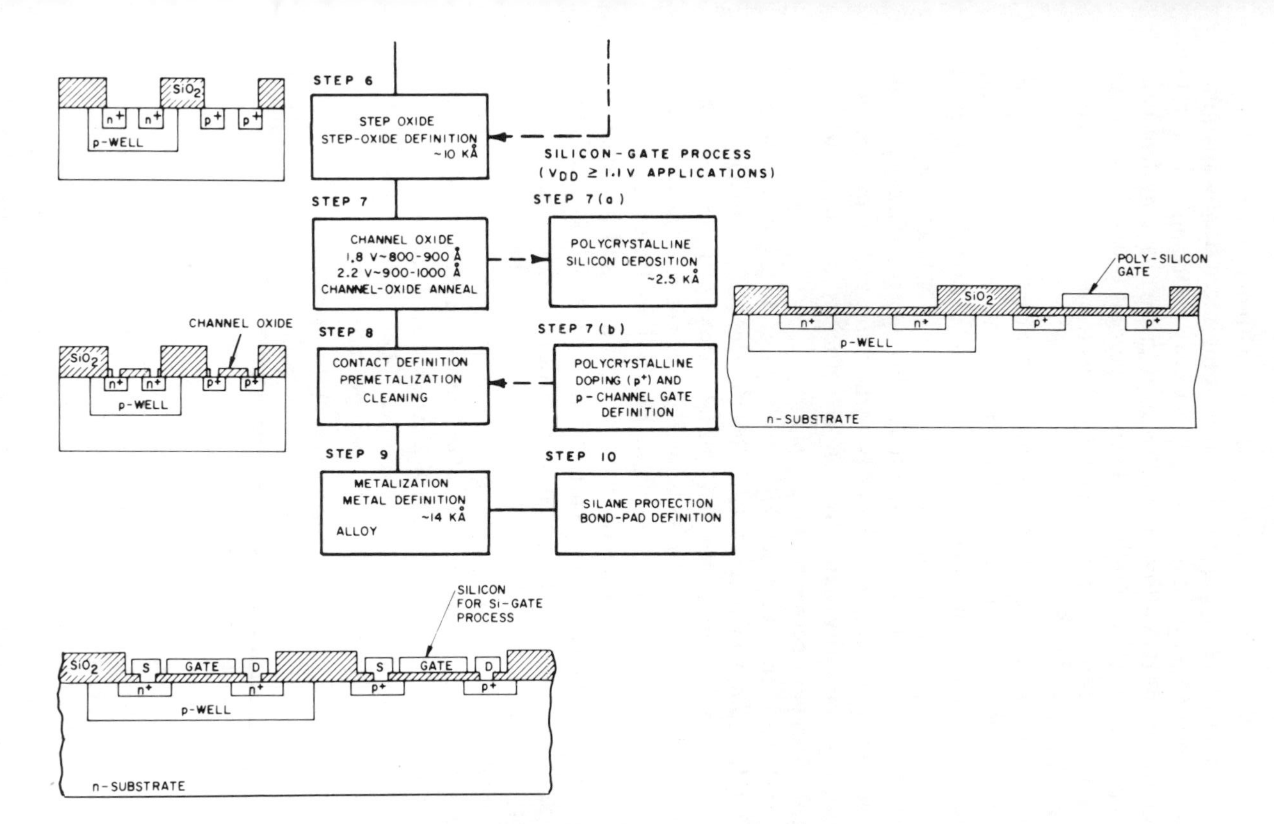

Fig. 4-11. Basic operations in the fabrication of COS/MOS integrated circuits. (Courtesy RCA Corp.)

as well as faster than core memories, and with microprocessors that allow data-processing capability to be built into virtually anything, the semiconductor industry is anxiously working on nMOS technology. nMOS allows 15 to 20 percent greater packing density of circuitry on a chip of silicon, 30 to 50 percent faster speed, lower voltage operation, and potentially lower cost than the more conventional pMOS.)

An even more recent development is the double-diffused metal-nitrate-oxide semiconductor (D-MNOS) transistor. This device has been proposed as a nonvolatile memory that has higher writing speed and lower writing voltage. In the fabrication process, phosphorus and boron are diffused consecutively through the same windows, which are photoengraved so as to form drain and source regions. Substrate impurity concentration is sufficient to obtain a fully depleted region between two p+ regions.

Perhaps the "hottest" MOS technology today is that termed CMOS—complementary MOS. CMOS, McMOS, and COS/MOS are all terms meaning the same thing—McMos being Motorola's CMOS line, and COS/MOS being RCA's. The reasons for CMOS being so popular are that it can be used in linear and digital applications over a wide variety of temperature and power-supply variations; it has the lowest current drain of any MOS device; it has high noise immunity; it dissipates very little power, for example, only 0.4 μW/bit at 10 volts for a 256-bit memory; and can be made compatible with TTL circuits without the need for additional interfacing circuits. A typical CMOS fabrication process is illustrated in Fig. 4-11. However, CMOS devices also lend themselves to fabrication by the silicon-on-sapphire (SOS) process, resulting in higher-speed devices with even lower power dissipation.

Figure 4-12 shows the stages in the p-channel SOS/MOS process. Beginning with an n-silicon sapphire wafer (A), a layer of SiO_2 is deposited over the entire silicon surface (B). Using standard photolithic techniques, this deposited SiO_2 is selectively removed from the regions where silicon is not required (C). The unwanted silicon can then be removed in a suitable etch, and the oxide mask stripped (D). The etch also leaves the silicon islands tapered, resulting in easier contacts and crossovers. The wafer is now ready for source-drain diffusions. A suitable diffusion source is a deposition of SiO_2 doped with boron, plus an undoped layer of SiO_2. The boron-doped SiO_2 is then selectively removed in the n-channel regions (E). The

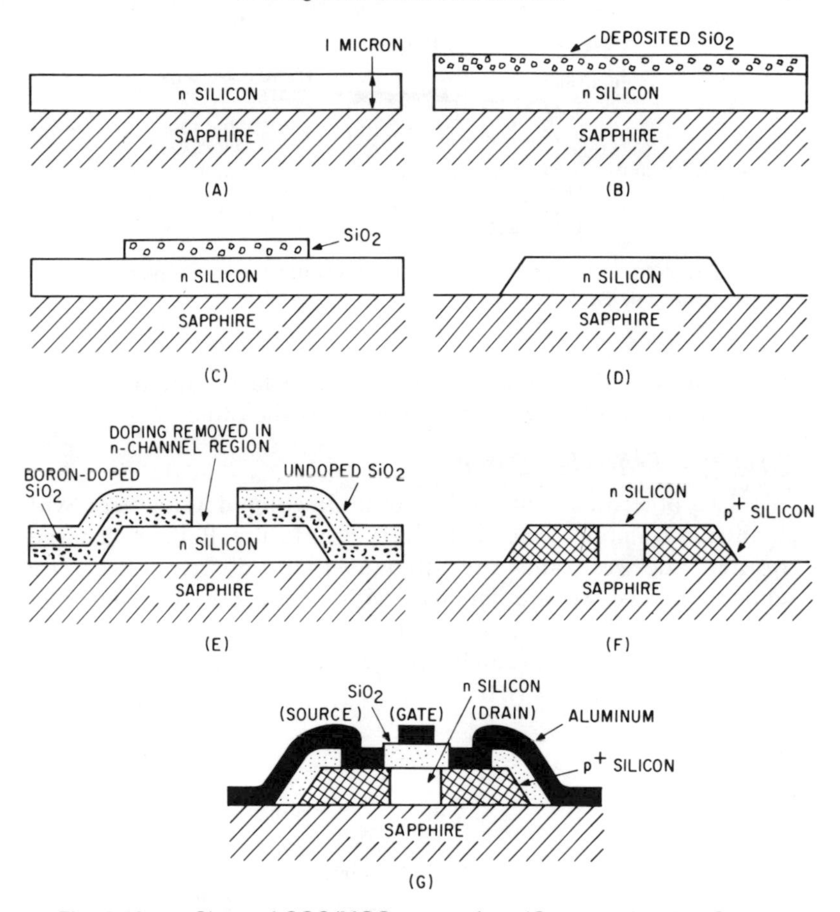

Fig. 4-12. p-Channel SOS/MOS processing. (Courtesy Inselek Corp.)

wafer is placed in a drive-in furnace, and the boron is diffused into the silicon. When diffusion is complete, the boron-doped oxides are stripped from the wafer (F). Finally, gate oxide is built up to the desired thickness, contact openings are etched, and aluminum connections are evaporated and defined (G). Because the silicon layer is so thin, SOS/MOS devices can operate at speeds comparable to those of bipolar devices and at voltages higher than 100V.

MOS capacitors can be formed using the SiO_2 insulating layer, as shown in Fig. 4-13. A metal deposit on the surface of the oxide

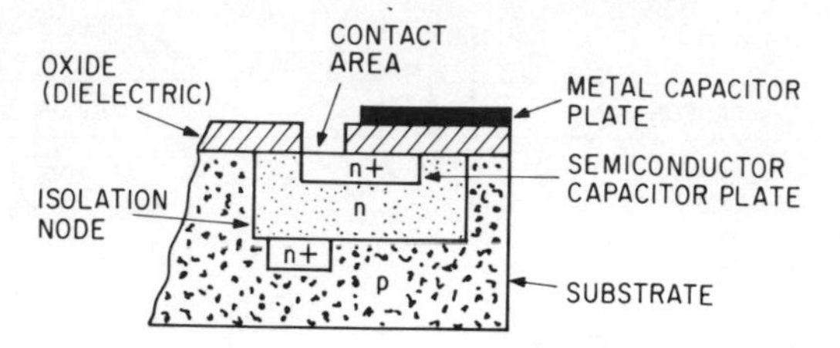

Fig. 4-13. Using the oxide as the dielectric for an IC capacitor. (Courtesy RCA Corp.)

forms one of the capacitor's conducting surfaces, and an n+-type region diffused into the n-type node forms the other.

Charge-Transfer Devices

Two other new types of IC are undergoing development, with significant results expected in the very near future. Both are "charge-transfer" devices, and are fundamentally different from the semiconductors we are familiar with. They use the controlled movement of electric charge to perform their functions and are generally classified as either charge-coupled devices (CCDs) or bucket-brigade devices (BBDs).

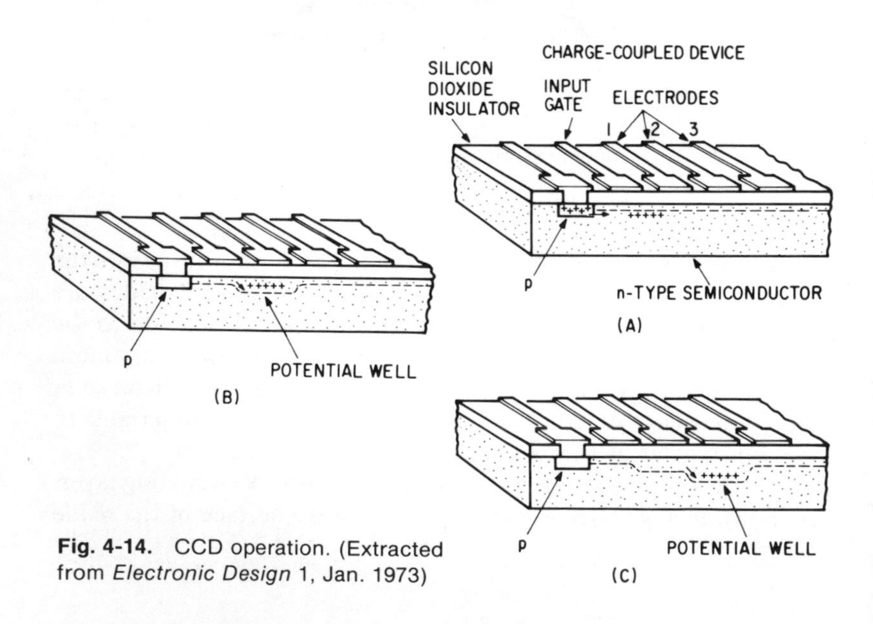

Fig. 4-14. CCD operation. (Extracted from *Electronic Design* 1, Jan. 1973)

CCDs consist of three layers: semiconductor, oxide, and metal electrodes, as shown in Fig. 4-14. There are several methods now being used to obtain the layering, as shown in Fig. 4-15. The first (A) is a simple three-phase device, but the close electrode spacings are difficult to produce. Two-level metalization CCDs, either two-phase or four-phase devices (B), eliminate spacing problems. Two-phase operation can also be obtained by using an ion-implanted-barrier CCD (C). An improvement over the implanted-barrier device is the conductively connected CCD (D) which also alleviates the spacing problem. The buried-channel CCD (E) eliminates the trapping effects of interface states, and promises a more efficient device. CCD operation requires that electrical charge in the form of minority carriers be transferred from the region under one electrode to the region under another. The charge represents information, and is moved by control of the voltage applied to the electrodes. When the information (in the form of a negative voltage) is applied to the input gate (Fig. 4-14), a channel in the n-type material opens up and draws

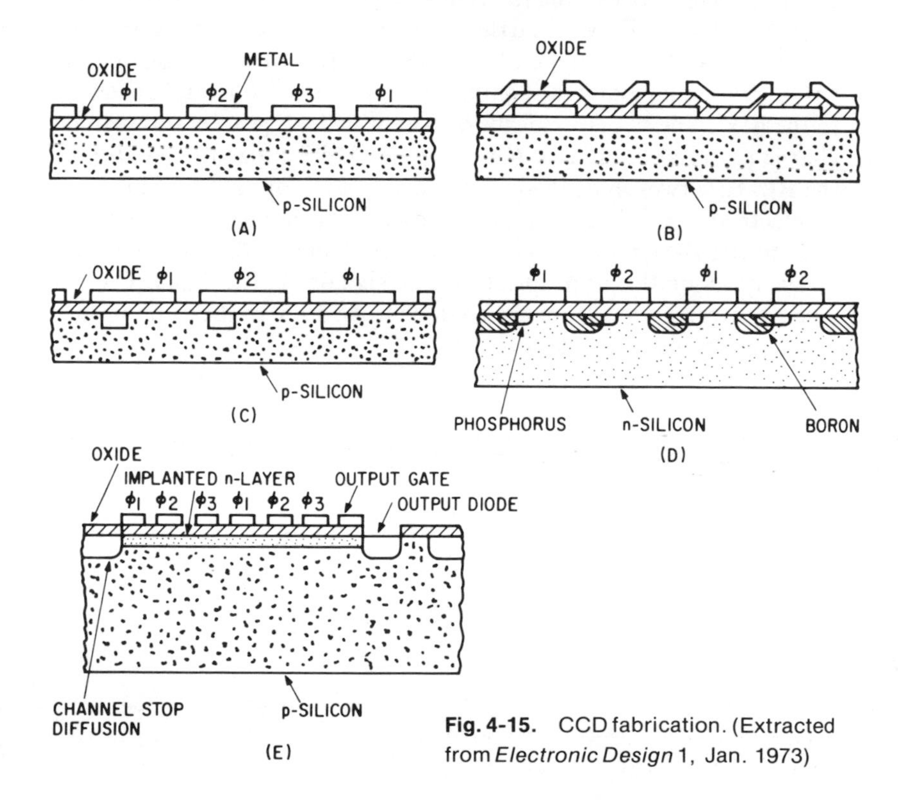

Fig. 4-15. CCD fabrication. (Extracted from *Electronic Design* 1, Jan. 1973)

positive charges from the input p-n junction to the region under the gate, as shown in A of the figure. When a negative voltage is applied to electrode 1, evenly distributed electrons in the n-type material are driven downward, away from the electrode, leaving a region of electron-depletion just below the oxide layer. This depletion region produces a potential well, as shown in B of Fig. 4-14. If the electrodes are spaced closely enough, positive charges passing through the channel under the input gate are drawn to electrode 1 and stored in the potential well.

The information (the positive charges) can be transferred from electrode 1 to electrode 2 by application of a negative voltage to electrode 2 and reduction of the voltage on electrode 1. This transfer voltage forms a deeper potential well under electrode 2, and the information flows to the well under electrode 2. Thus, by proper manipulation of electrode voltage, information can be transferred, as shown in C of Fig. 4-14.

BBDs differ from CCDs in that they can also be constructed from discrete devices. Information is stored as majority carriers and transferred as minority carriers. Storage occurs in an array of capacitors as "charge deficit." Charge-transfer circuits consisting of MOS transistors allow the charge deficit to move from one capacitor to the next. The structure of the bucket brigade, Fig. 4-16, shows that charge is stored in offset p-regions under MOS capacitors, and that the BBD is a two-phase device. When a negative voltage is applied to an electrode, the diffused area underneath it becomes reverse biased and inverts the channel between this diffusion and the next one. By application of the appropriate clock signals to the BBD, excess charge is transferred from the source of one FET to the drain of the next via the inverted channel.

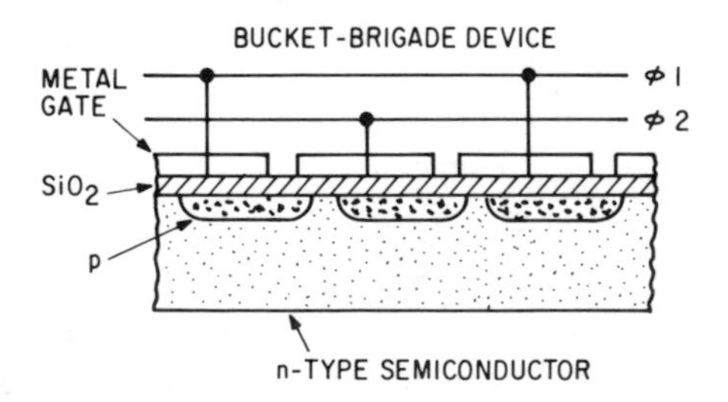

Fig. 4-16. BBD operation/fabrication. (Extracted from *Electronic Design* 1, Jan. 1973)

Bubble Memories

Bubble memories do not fall strictly within the classification of integrated circuits. However, the bubble memory, which was conceived by the Bell Telephone Laboratories, is an IC-like device composed of garnet material in which data are stored as very small "bubbles." Physically, a "bubble" device consists of a thin-film magnetic garnet grown epitaxially on a nonmagnetic garnet substrate. With the application of an all-enveloping magnetic field, the garnet film reacts somewhat like magnetic tape during recording—the magnetic field sets up extremely small stable areas, or "domains," of reverse magnetization. Although these domains appear to be bubble-like when viewed in polarized light, in actuality they are very small cylinders that can be forced at high speeds along preset paths past some form of sensor. Their rapid growth in popularity is attributable to their storage capacity. For example, a 0.3-in.^{2} of thin-film garnet has the capacity for 100,000 bubbles, each containing stored data. Also, a new compound, bismuth-thulium garnet developed by RCA, promises to increase the readout speed by a factor of 100, making bubble memories even more desirable for computer and other data-processing systems.

Metalization

Once the desired elements have been formed, both metal contacts to the elements for use by the outside world, and metal element-interconnection paths must be added. This metalization, which uses aluminum most frequently as the metal, comprises three facets: metal deposition, interconnection pattern delineation, and alloying.

Conventional metal deposition is formed by placing the processed wafer in a vacuum chamber. The source material, also in the vacuum chamber, is then heated above its vaporization point. As a result, metal is released by vaporization and condensed on the IC, coating the latter to a controlled thickness. The metal is then subjected to a process whereby photoresist is added, exposed to an ultraviolet light through a photomask, and the unwanted metal then acid-etched away. The results are the contacts shown in Fig. 4-17 and the interconnections shown in Fig. 4-18. To assure good electrical contact between the metalization and the IC elements, the wafer is again heated in a furnace for a controlled period of time, so that the metal alloys slightly with the semiconductor junctions.

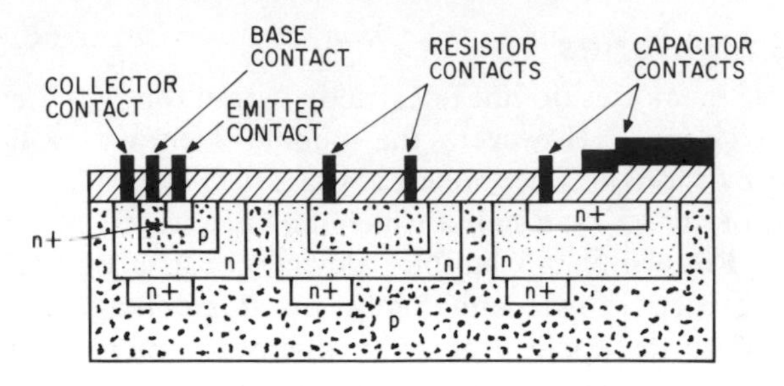

Fig. 4-17. Adding the metalized contacts to the IC elements. (Courtesy RCA Corp.)

Once metalization is complete, the wafer is then cut up into the individual IC chips. These latter must then be connected to their respective package leads. Traditionally, the connections have been via thermocompression or ultrasonic bonding methods. However, there is now a strong industry movement toward the use of interconnect patterns, which substitute for the traditional wire bonds, supported on a flexible backing material which resembles movie film. Sprocket holes in the film are used for the indexing and alignment operations. With this new technique, high-speed devices

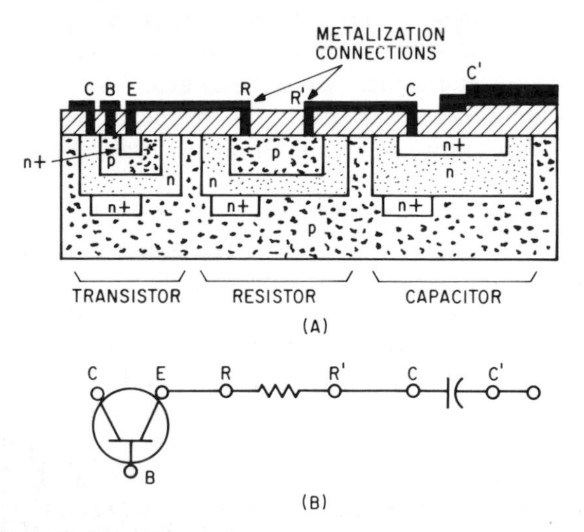

Fig. 4-18. A complete silicon chip containing a transistor, resistor, and capacitor (A), and the circuit diagram of the interconnected elements (B). (Courtesy RCA Corp.)

will be aided because of the short lead lengths that can be obtained. Also, large packages with long lengths can be replaced by interconnecting the chips directly to ceramic substrates.

Encapsulation

The role of moisture in the performance of integrated circuits has become increasingly important as the circuits have become more sophisticated and as they have to meet tighter performance specifications. Water vapor can adsorb on the oxide or nitride layers that form the surfaces of most ICs. If enough water is adsorbed, leakage currents can develop, resulting in device failure. The main purpose of encapsulation, therefore, is to provide protection against moisture, as well as to provide mechanical protection.

The choice of encapsulant depends largely on the type of IC being encapsulated. Thus, a silicon IC would permit the use of either silicone resin or silicone rubber. A film (generally, tantalum-nitride) IC, on the other hand, cannot withstand the high temperatures required to cure a silicone-resin encapsulant. So, low-temperature-curing silicone rubber is the solution. As for hybrids, they can contain all sorts of ICs and discrete components, consequently resulting in the preparation of a compromise encapsulant.

A typical encapsulation process begins with a mask and a frame being placed over a batch of ICs. The mask is "windowed" so that each window corresponds to one IC in the batch. The mask is then sprayed with a mixture of silicone resin and xylene, similarly to spraying paint through a stencil. The xylene evaporates, leaving the ICs coated with a thin layer of the silicone resin. The entire batch is then placed in an oven to accomplish the necessary curing.

Dielectric Isolation

The monolithic ICs discussed so far have one important feature that distinguishes them from films or discrete components: reverse-biased p-n junctions are employed to obtain isolation between various circuit parts. Such junctions involve the following:

1. Parasitic capacitances that degrade performance at high frequencies.
2. Reverse leakage currents at the junctions that are highly troublesome in high-radiation environments.
3. Limited junction-breakdown voltages.

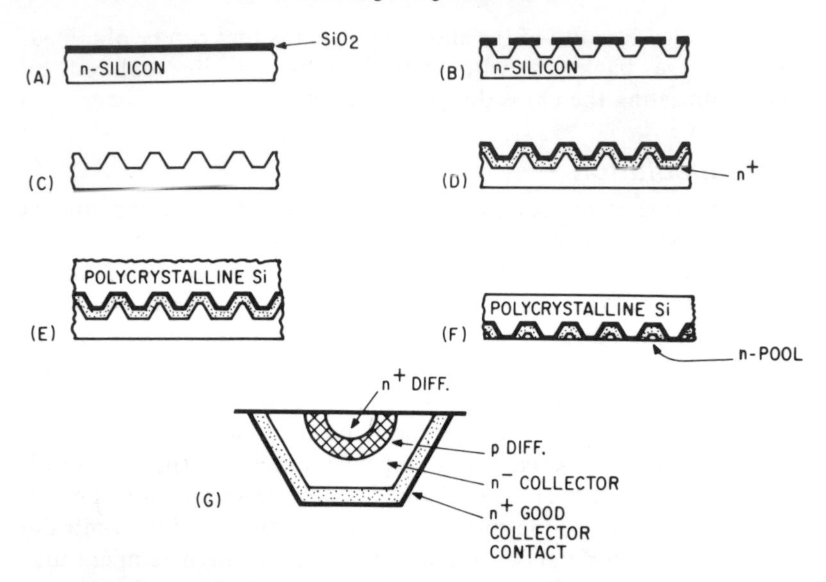

Fig. 4-19. The dielectric-isolation process. (A) Cross-section of starting material. (B) Etch for isolation pool (moat). (C) Lap. (D) N+epitaxial deposition and oxidation. (E) EPI—deposition of polycrystalline silicon. (F) Finished dielectrially-isolated wafer through backlap and surface oxidation. (G) Blown-up section of one pocket (wafer turned around) with additional diffusions to make n-p-n transistor.

Figure 4-19 shows the fabrication process for dielectrically isolated structures. For simplicity, only a transistor is shown. If desired, resistors or MOS capacitors can be diffused into their own isolation pools. However, for very high reliability, compatible thin films are usually deposited on top of the SiO_2 passivation layer.

Although the processing is more complicated than for monolithic ICs, dielectric isolation practically eliminates inter-component parasitics. Further, if the compatible thin-film technique is used, it carries along its superior temperature coefficients and resistance accuracy.

Beam-Lead Integrated Circuits

Beam lead identifies a structure in which gold beam leads are extended over the chip edges as cantilever beams. Figure 4-20A shows the construction of a beam-lead, sealed-junction transistor.

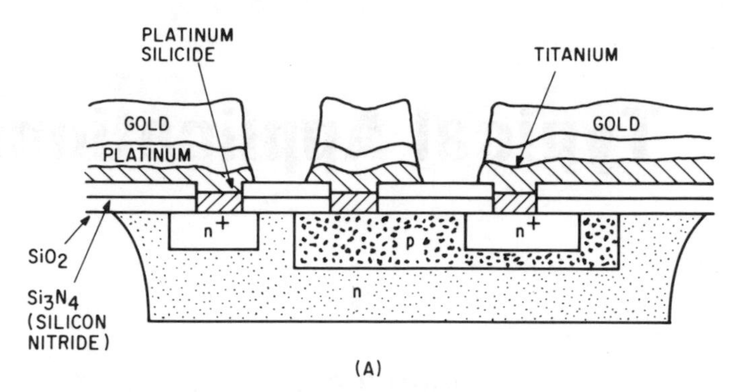

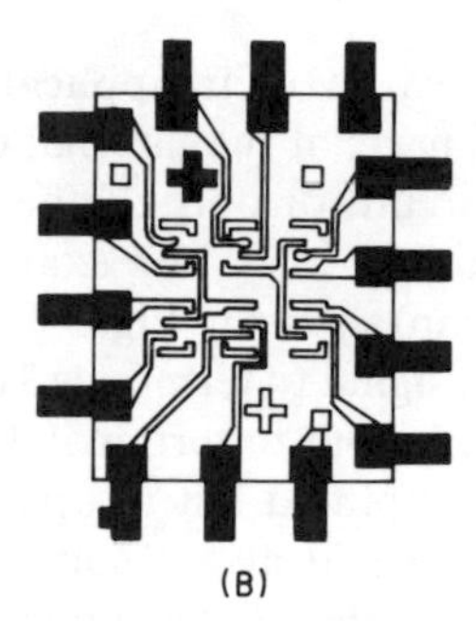

Fig. 4-20. Beam-lead constructions. (A) A beam-leaded, sealed-junction IC transistor. (B) CA3054L beam-lead, dual independent differential amplifiers. (Courtesy RCA Corp.)

"Sealed junction" connotes that the transistor junctions are sealed by a passivating layer, silicon nitride (Si_3N_4), for example. Such passivation both protects the junctions from sodium ions, a ubiquitous contaminant that easily penetrates SiO_2, and eliminates the need for hermetic sealing. After etching, platinum silicide electrodes are formed by platinum deposition. A three-metal system is used to form a strong reliable bond that adheres well to the silicon-nitride surface. Titanium and platinum are first sputtered on, with the gold leads then formed by plating. A complete beam-lead IC, containing dual independent differential amplifiers is shown in Fig. 4-20B.

Beam-lead devices eliminate the need for reverse-biased junctions as the isolating mediums between individual circuit elements. Also, individual circuits can be air-isolated, or embedded in a ceramic matrix, and so provide dielectric isolation.

5 Typical Applications

Introduction

Most ICs offer considerable flexibility in their application. Because of the many terminals at various parts of the internal configuration one can use either the entire circuit, or only the needed portions, with the unused portion remaining idle. For example, examine the IC operational amplifier (op amp) shown in Fig. 5-1. The entire IC may be used by applying an input signal to terminals 3 or 4, and picking the output off terminal 12. An input to terminal 3 will result in an output at terminal 12 that is amplified but of opposite polarity (inverted). An input to terminal 4 will also result in an amplified output, but with the same polarity as that of the input (non-inverted). The IC thus offers two separate identical-gain amplifiers, with a common low-impedance (emitter-follower) output.

Unused Inputs

Sometimes, the whole circuit isn't required. For example, you can apply an input to terminal 11, and use only Q_{10} as an emitter-follower output. Or, you can put an input signal on terminal 9, thereby using Q_4 as a common-emitter amplifier direct-coupled through Q_8 to output emitter-follower Q_{10}. Further you can use a single input stage as a common-emitter amplifier by applying the input signal to terminal 3 and taking the output from terminal 1; or by going in on terminal 4 and taking off at terminal 9.

One must be careful, however, as to the device being used when not all inputs are used. In a CMOS device, all input pins must be connected to a voltage level between V_{SS} (the most negative potential) and V_{DD} (the most positive). In circuit configurations in which input pins connect directly to NAND gates, the unused inputs must be tied to V_{DD} (high-level state) or tied together with another input pin to a signal source. Either of these connections is necessary, because

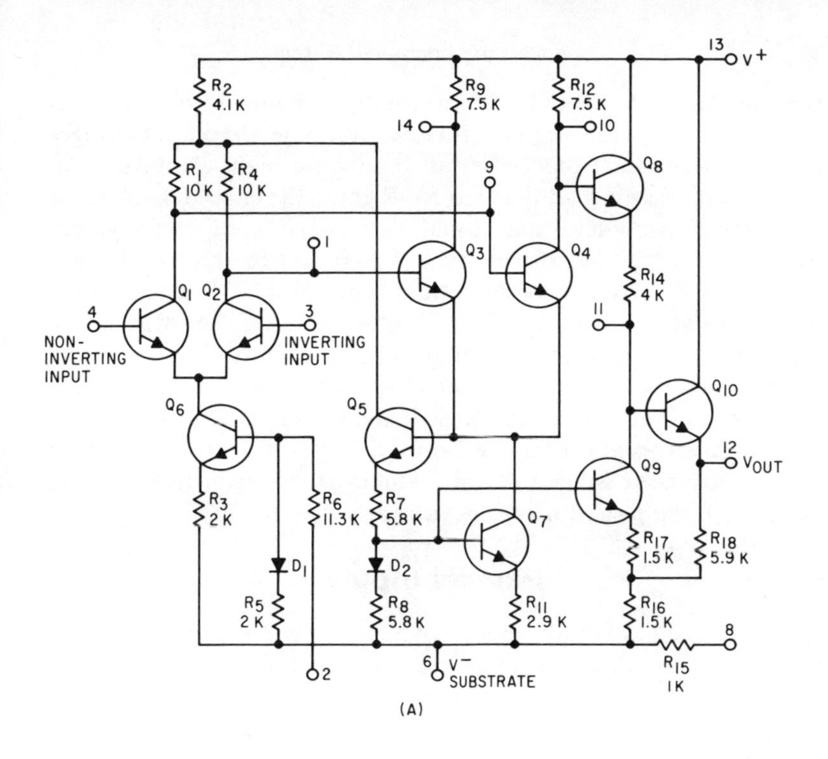

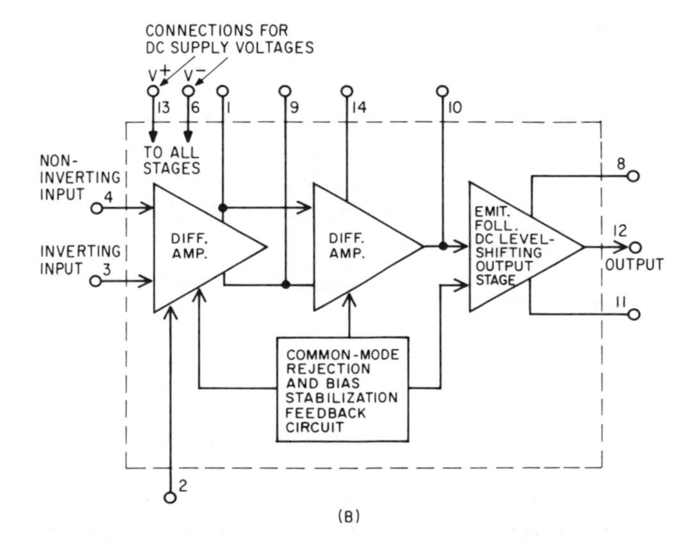

Fig. 5-1. RCA CA3030 Operational Amplifier. (A) Schematic. (B) Functional block. (Courtesy RCA Corp.)

the output of any NAND gate normally remains in the high-level state for as long as any input is in a low-level state; the output changes to a low-level state only when all inputs are high. Conversely, in circuits which connect directly to NOR gates, the unused inputs must be tied to V_{SS} (low-level state, usually ground) or connected together to another input pin driven by a signal source. Either of these connections is required because the output of a NOR gate is in the low-level state if any input is in the high-level state. The output of the NOR gate goes high only when all inputs are low. Floating inputs (unused inputs) guarantee neither a logic 0 nor a logic 1 condition at the output of the device, but cause increased susceptibility to circuit noise and can result in excessive power dissipation. Floating inputs to high-impedance CMOS gates can result in linear-region noise biasing when both the p- and n-type devices are ON.

Parallel Inputs

The inputs of multi-input NAND and NOR gates are sometimes wired together and connected to a common source. In the case of NAND gates, where as many as four input pins may be wired together a slight increase in speed occurs when more than one input is tied to the same signal. More importantly, however, the output source current of the device is increased proportionately to the number of inputs wired together. When the inputs of a NOR gate are tied together to a common input signal, the gate experiences a higher sink current and a slight increase in speed. The speed increase in both NAND and NOR gates results from the lower ON resistance of the paralleled devices. The increase in speed is minimized by a compensating speed decrease caused by the added capacitance of the driving source as well as capacitance internal to the device itself.

Both source- and sink-output currents are increased by paralleling two or more similar devices on the same chip. This increased drive capability also increases speed if the increase in capacitance loading is not excessive. When devices are paralleled, power dissipation also increases.

Differential Amplifier

The differential op amp, as exemplified by the CA3030 IC, is one of the most useful linear ICs. The type of full-IC operation discussed earlier is illustrated in Fig. 5-2A; with no overall feedback, the circuit

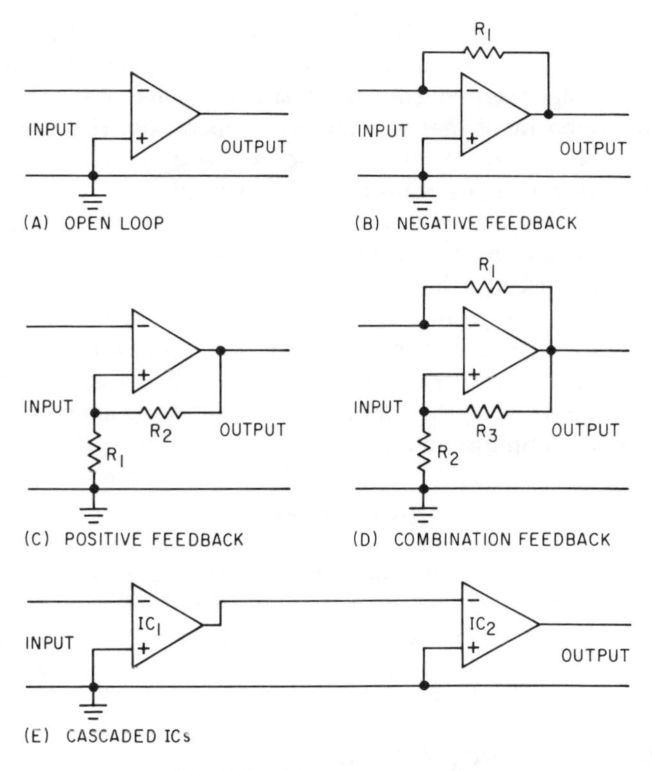

Fig. 5-2. IC op amp operation.

realizes the open-loop gain voltage of the IC. On the other hand, overall feedback may be used. For example, we have negative feedback provided by R_1 in Fig. 5-2B; negative because the feedback is applied to the inverting input terminal of the IC. Figure 5-2C shows a method of providing positive feedback, through voltage divider R_1-R_2. (If this feedback is made high enough, the op amp can become an oscillator.) Lastly, both positive and negative feedback are shown in Fig. 5-2D.

Although op amps usually provide high gain, the gain of a single IC may not be enough. In such instances, two or more IC op amps may be "cascaded," as shown in Fig. 5-2E. In cascaded ICs, feedback, if required, can be applied to the individual ICs or the cascade itself. In cascades, individual and combined phase shifts must be considered when feedback paths are planned. For example, with the cascade shown in Fig. 5-2E, a signal applied to the negative input terminal of IC_1 will appear *uninverted* at the output of IC_2, and vice versa if the signal is applied to the positive terminal of IC_1. Thus, we have the opposite of that with single-IC operation. However, if we should add a third IC to the cascade, we'd revert to single-IC operation as far as signal inversion is involved.

The remainder of the chapter shows typical IC applications, using both single and multiple ICs. Included are both linear (analog) and digital circuits, but by no means the full gamut of IC potentialities. For example, linear ICs are often used in digital applications, such as putting to use an op amp as a multivibrator. Also, digital ICs are frequently employed in linear applications, a typical example being the use of a three-input AND gate as a microphone mixer-amplifier in a hi-fi rig.

Linear Circuits

One of the most common uses of linear ICs is as an amplifier. Consider the possibilities: operation at dc up to MHz ranges; voltage gains on the order of 100 dB or more; single-ended or balanced output; voltage or power output; inversion or noninversion; high-to-low impedance conversion; etc.

Audio-Frequency Amplifiers

One of the more simple applications is as an audio-frequency amplifier. Take, for example, the circuit illustrated in Fig. 5-3B. Here, only one external resistor is used.

The CA3000 contains two direct-coupled amplifier stages, each consisting of an input emitter follower plus an output common emitter. In the connection shown in Fig. 5-3, only one of these amplifiers is used; the other floats. The voltage gain of this connection is 30 dB. Without any external compensating components, the frequency response is dc to 1 MHz. Input impedance is 0.1 megohm, and output impedance is 8k. Both amplifiers may be used in cascade, with corresponding increase in gain, by connecting an interstage coupling capacitor externally between IC terminals 10 and 6 and taking the output from terminal 8, instead of terminal 10.

Gain control (by means of a potentiometer) and dc isolation of input and output circuits (by means of blocking capacitors) may readily be achieved, if required, through the connection of external components.

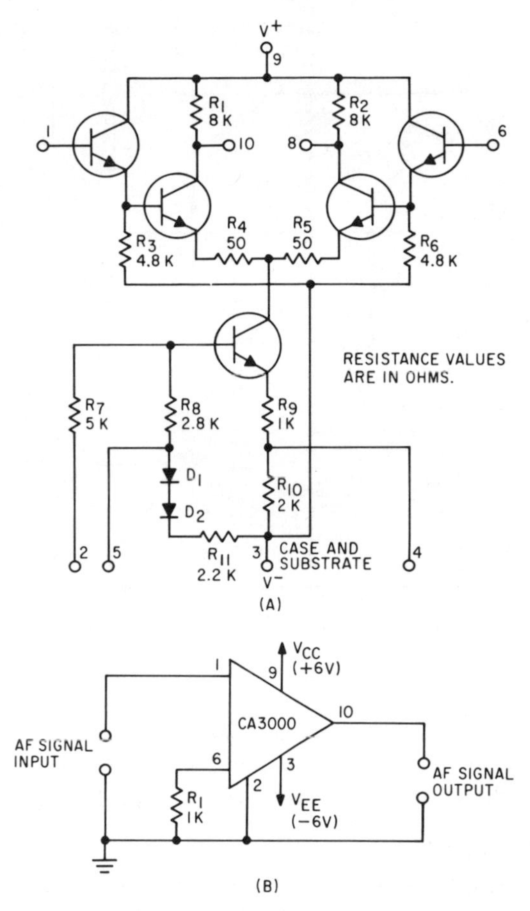

Fig. 5-3. Dc amplifier used as an audio-frequency amplifier. (A) CA3000 schematic. (B) Audio-frequency amplifier configuration. (Courtesy RCA Corp.)

Audio-Power Amplifier

A 1 W audio-power amplifier is shown in Fig. 5-4. Here the input configuration can readily be recognized as a fairly conventional differential amplifier, with Q_1 serving as the constant-current source for the differential transistor pair Q_2 and Q_3. The function of Q_4 is to balance the dc collector voltages of Q_2 and Q_3.

Q_6 and Q_7 are the common-collector power drivers for the modified totem-pole output arrangement comprising Q_8 and Q_9. Q_{5A}

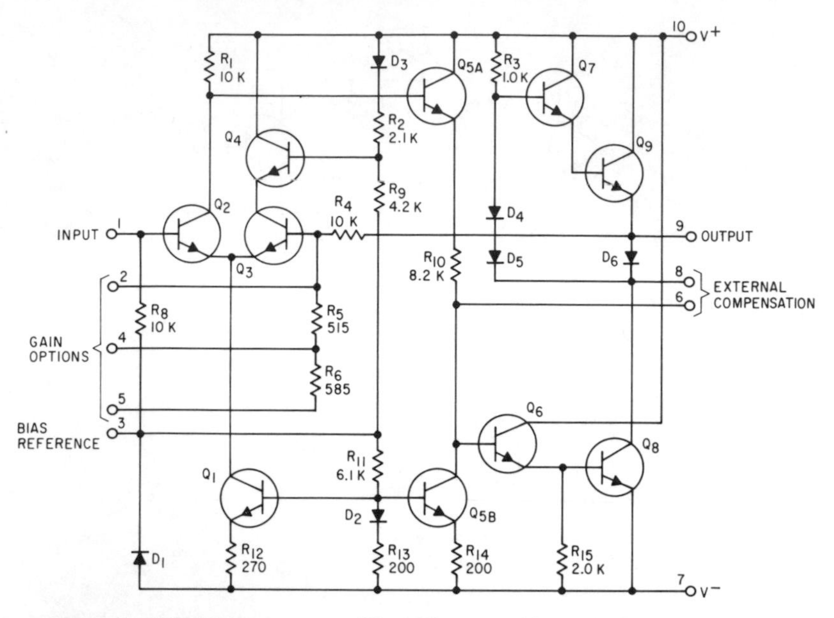

Fig. 5-4. MC1554 power amplifier. (Courtesy Motorola Semiconductor)

is the signal driver for Q_8 and Q_9, while Q_{5B} provides a constant-current source that, in combination with R_{10}, serves as a level translator to direct couple from R_1 to the input of Q_6.

Video Amplifier

The extended bandwidth of many linear ICs suits them for video-amplifier circuits. Figure 5-5B illustrates one such configuration. Driven by a 50-ohm source, this RCA CA3001 amplifier provides a voltage gain of greater than 15 dB at frequencies ranging from 0(dc) to 5 MHz. The circuit shown in Fig. 5-5A consists of a differential-amplifier pair comprising Q_3 and Q_5, the current of which is controlled by constant current transistor Q_4. Q_1, Q_2, Q_6, and Q_7 are operated in a common-collector configuration to provide a high-impedance input and low-impedance output. The high-frequency response of the circuit is determined mostly by the capacitance and resistance in the collectors of Q_3 and Q_5.

Automatic Fine Tuning

An application of a linear IC is shown in Fig. 5-6B, illustrating the use of an RCA CA3044 in an automatic fine tuning application.

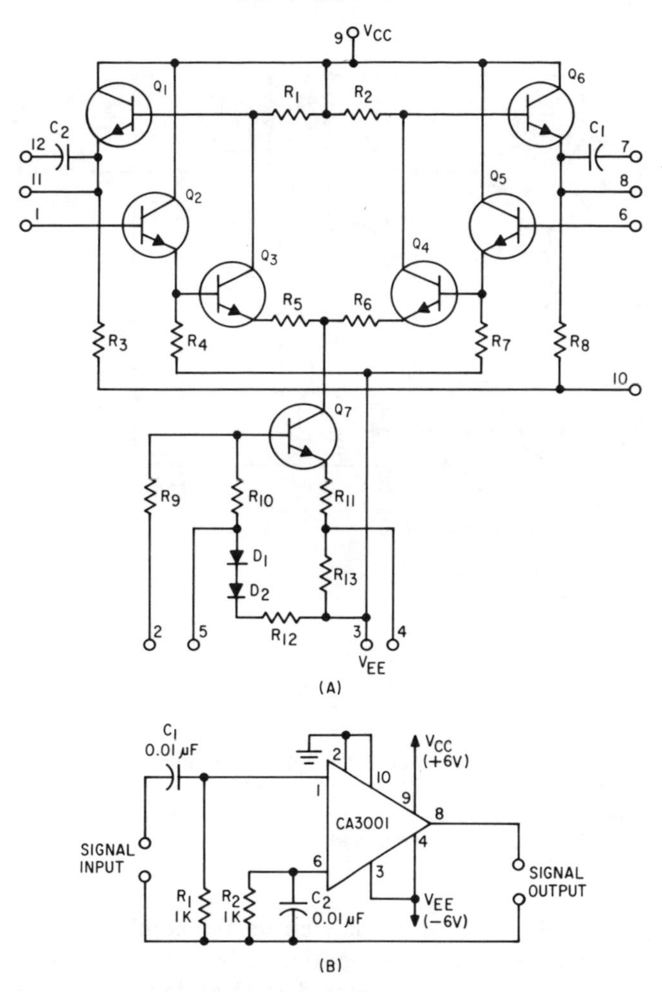

Fig. 5-5. Wideband IC amplifier used as video amplifier. (A) IC schematic diagram. (B) Video-amplifier configuration. (Courtesy RCA Corp.)

In this system, the IC provides all of the sigr components needed (with the exception of the tunec transformer) to derive the AFT correction signals f the video-IF amplifier. The other components of signal coupling and power-supply decoupling a signal processing in the video-IF range.

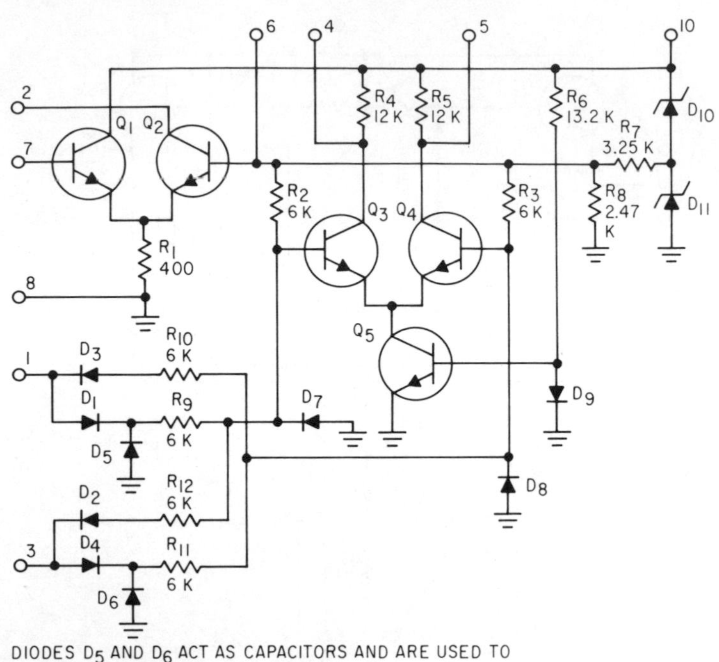

Fig. 5-6. System diagram of a typical automatic-fine-tuning (AFT) application showing the CA3044 in use in a color TV receiver. (A) Schematic diagram of IC; all resistance values in ohms unless otherwise specified.

The CA3044 can be considered as the combination of four functional blocks: a limiter-amplifier, a balanced detector, a differential amplifier, and a regulator. The 45-MHz limiter-amplifier is basically a differential amplifier that supplies a peak-to-peak output current of approximately 4 mA for input levels above the limiting threshold. The use of a load impedance which does not exceed 2000 ohms eliminates detuning effects caused by saturation of the mplifier under worst-case conditions. In the system shown, the load npedance at the center frequency is about 1800 ohms.

The diode matrix composed of D_1, D_2, D_3, and D_4 constitutes a anced detector that converts the output of the phase transformer filtered dc signal. Diodes D_1 through D_4 perform the detection ion. Diodes D_7 and D_8 are always reverse-biased and serve as

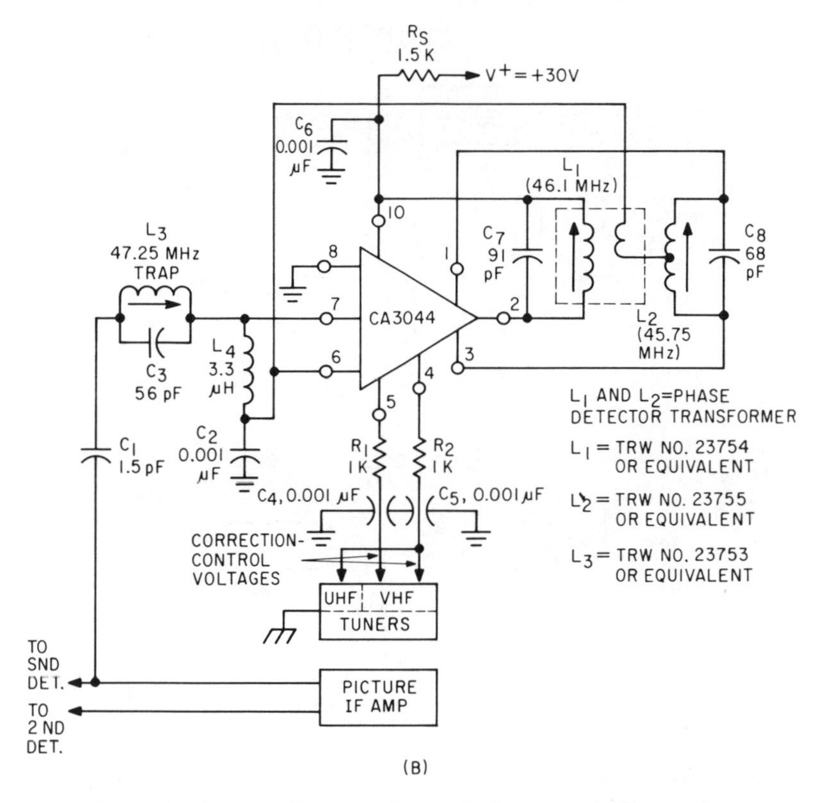

Fig. 5-6. System diagram of a typical automatic-fine-tuning (AFT) application showing the CA3044 in use in a color TV receiver. (B) Automatic-fine-tuning network. (Courtesy RCA Corp.)

capacitors; they filter the output of the detector in conjunction with resistors R_9 through R_{12}. Diodes D_5 and D_6 are included to balance the parasitic diodes that exist between the cathodes of D_2 and D_3 and the substrate.

Transistors Q_3, Q_4, and Q_5 form a constant-current driven differential amplifier that is directly coupled to the output of the detector. The amplifier provides sufficient power to allow the use of a low-cost tuning element. The output impedance at either output of the amplifier is approximately 12,000 ohms.

The zener-diode regulator, comprising D_{10} and D_{11}, provides the regulation necessary for a differential post-detection amplifier that is

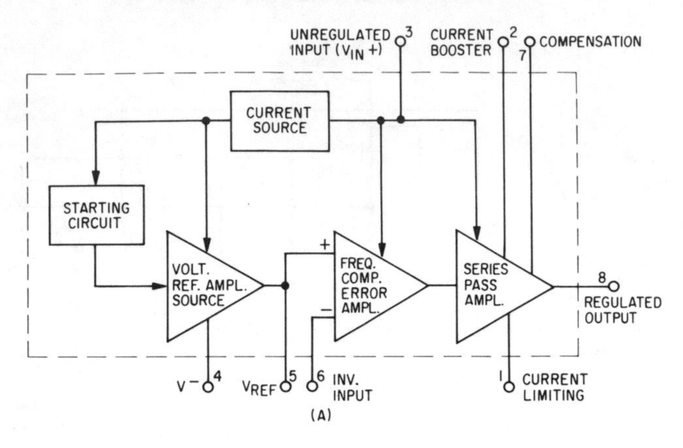

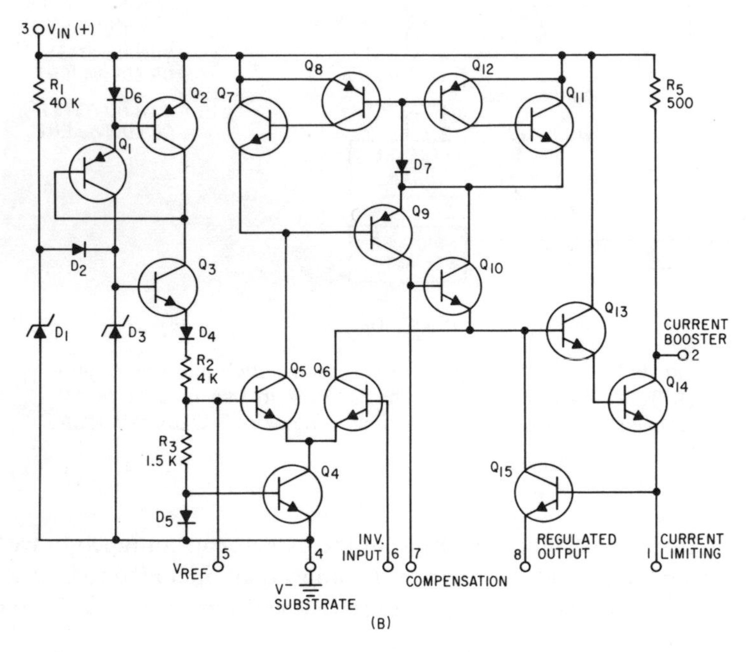

Fig. 5-7. System diagram of the CA3055 integrated-circuit voltage regulator. (A) Voltage-regulator configuration. (B) Schematic diagram of IC; all resistance values in ohms unless otherwise specified. (Courtesy RCA Corp.)

both stable and independent of temperature and power-supply variations.

During normal operation, the proper dc bias for terminals 1, 3, and 7 is supplied through terminal 6 and external RF coils. RF bypassing is required both for terminal 6 and for terminal 10, which is connected through the primary winding of the detector transformer to terminal 2.

Voltage Regulation

Lastly, the IC has greatly simplified the design and construction of voltage-regulated dc power supplies by providing the dc amplifier required for control. Still further simplification has been provided by special ICs that include the entire voltage regulator circuit, thus requiring no outboard components. Figure 5-7 illustrates the CA3055 IC, designed specifically for service as a voltage regulator. It can deliver output currents up to 100 mA, at output voltages from 7.5 to 40V, over a temperature range of -55° C to +125° C. It can be used as a shunt voltage regulator, and a high-current regulator. A typical power-supply application utilizing this IC is given in Fig. 5-8.

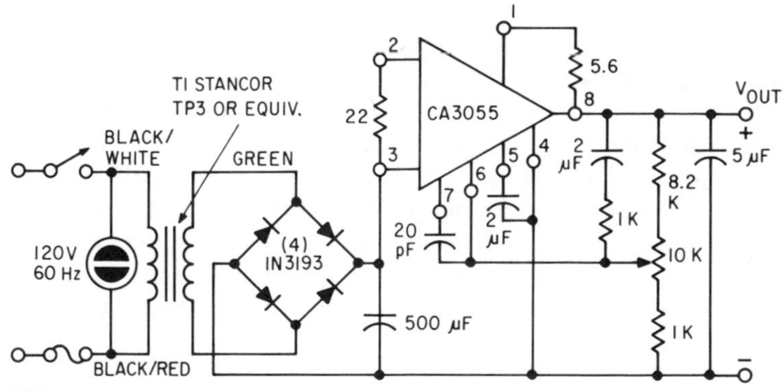

Fig. 5-8. Application of the CA3055 in a typical power supply; all resistance values in ohms unless otherwise specified. (Courtesy RCA Corp.)

Digital Circuits

Chapter 2 has already presented the basic building blocks of digital circuits. The remainder of this chapter describes some typical applications of those blocks.

Multivibrator

Figure 5-9 shows a Transitron SUHL TG90 IC used to make up a one-shot multivibrator. The IC itself consists of three 2-input AND gates that are ORed together, then inverted.

The trigger is a standard SUHL input, +5V, and the remaining inputs can be used to gate the trigger (read and write commands). The positive-going edge of the input pulse turns transistor Q_1 off while forcing the output of the inverter gate high to present the leading edge of the output pulse. The output of Q_1 is fed back to the other side of the Exclusive-OR gate. When the collector voltage of Q_1 reaches the Logic "1" threshold of the gate it locks in the output condition established by the trigger pulse. The high output level of Q_1 will persist until the capacitor has charged up to the E-B threshold of Q_1 turning it back on. The output pulse can never be shorter than the input pulse and the minimum input pulse width is equal to the T_{off} of Q_1 plus $T_{d_{on}}$ of the Exclusive-OR gate (typically 30 nsec).

The repetition rate and duty cycle are related to the value of R_1 used. To maintain a constant output pulse width at higher frequencies the potential at the junction of R_1 and C must recover to V_{CC} before the circuit is again triggered. The cascode output of the gate supplies charging current through a 120-ohm pull-up resistor during the initial recovery phase. When this potential reaches 3.5V, the cascode starts to turn off requiring R_1 to pull this point up to V_{CC}. A lower value of R_1 will then decrease the recovery time allowing for higher frequency operation, while raising the total power consumption and demanding a higher fan-out gate.

Shift Registers

It was pointed out in Chapter 2 that two simple flip-flops are necessary to store one bit of information. This is illustrated in the two-phase shift register shown in Fig. 5-10. Each stage comprises two NOR gates feeding a flip-flop, with a storage and transfer of one bit requiring two stages. When shift signal X goes negative, data from the input is loaded into stage A1, and data from A2 is transferred into B1. During this time signal Y is positive, so the contents of stages A2 and

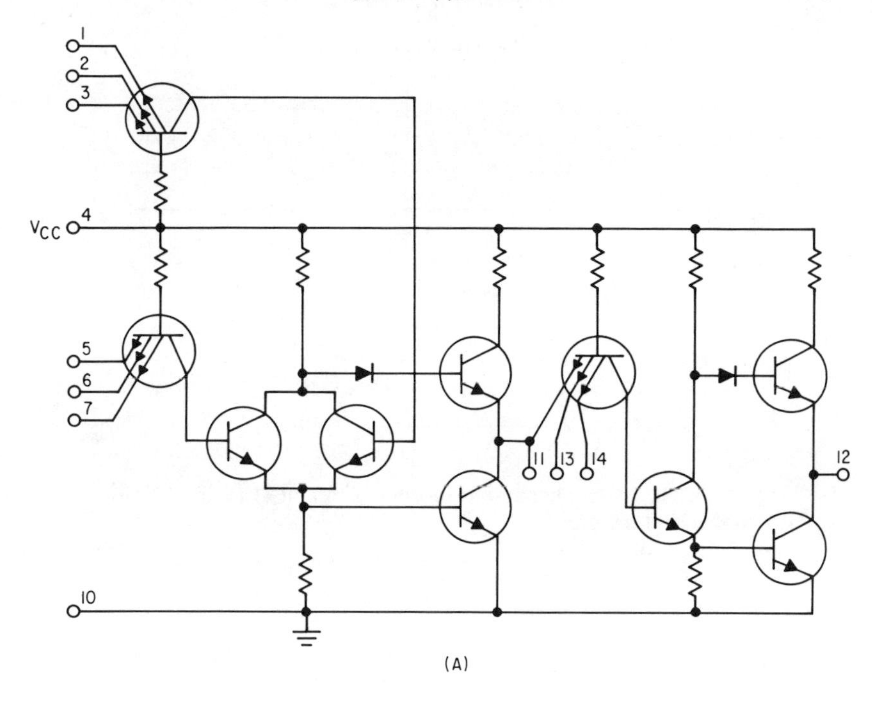

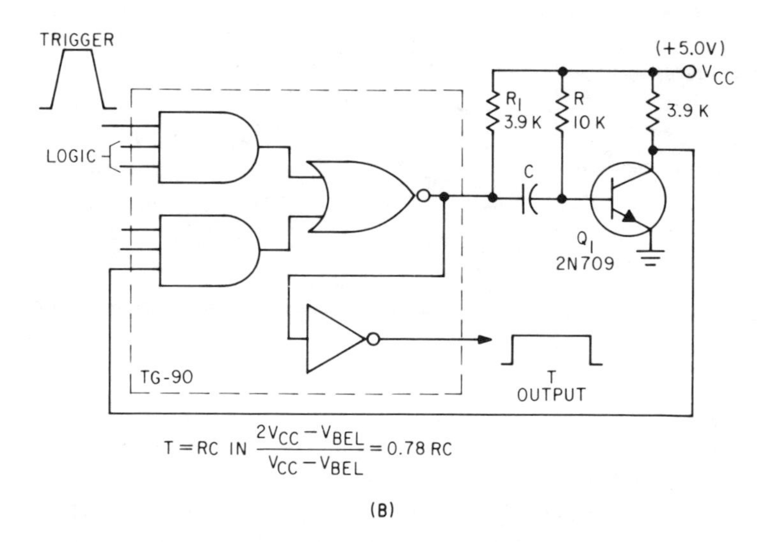

Fig. 5-9. Typical one-shot multivibrator using a TG90 Exclusive-OR gate. (A) TG90 schematic. (B) One-shot configuration. (Courtesy Transitron Electronic Corp.)

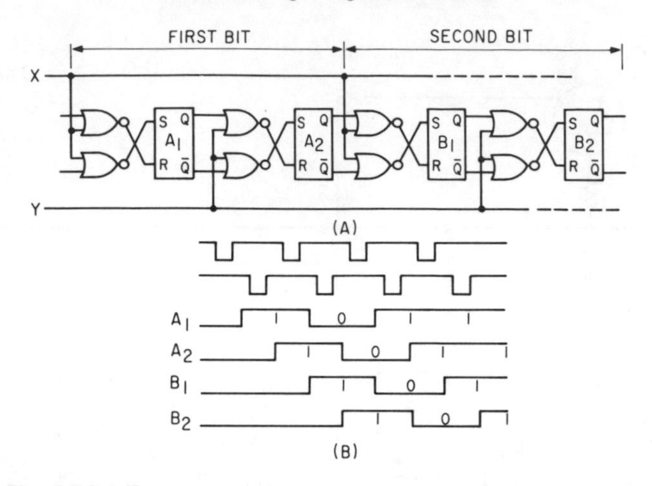

Fig. 5-10. Two-phase shift register using NOR logic, RS flip-flops, and waveforms.

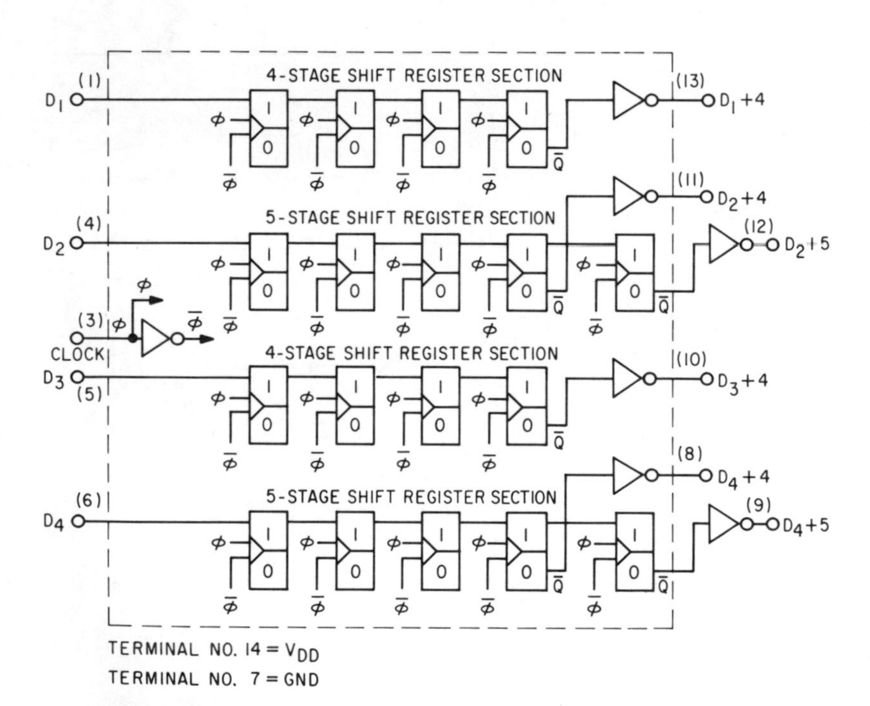

Fig. 5-11. Logic diagram of the CD4006A, eighteen-stage static shift register. (Courtesy RCA Corp.)

B2 are held steady for sampling by shift signal X. When signal Y goes negative, A1 transfers into A2 and B1 transfers into B2. As the waveforms show, the pattern of bits "shifts" down the register as each successive shift pulse gates the data into the successive stages.

Figure 5-11 illustrates how the simple two-stage shift register of Fig. 5-10 has graduated. Here we have an eighteen-stage shift register, the CD4006A IC. Each register stage shown comprises a static "master-slave" flip-flop, which is simply a configuration of two flip-flops wherein the "master" or input flip-flop samples the incoming data signals, while the "slave" or output flip-flop holds the data from a previous input, and then shifts it out when the master flip-flop is triggered again.

The CD4006A comprises four separate shift-register sections, 2 four-stage sections, and 2 five-stage sections. Each register section has independent data, "D," inputs to the first stage. The clock input is common to all eighteen stages. Through appropriate connection of inputs and outputs, multiple-register sections of 4, 5, 8, and 9 stages, or single-register sections of 10, 12, 13, 14, 16, 17, and 18 stages can be implemented with this single IC.

Counters

The "ripple carry" counter is another application of the digital building blocks. In the ripple carry counter, the input to be counted is entered into a trigger input in the first stage. The first stage output then triggers the second stage which in turn, triggers the third stage, and so on. This type of counter is frequently used in instrumentation.

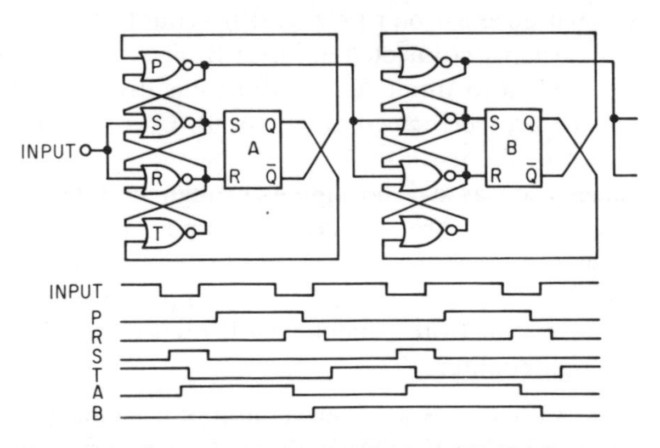

Fig. 5-12. Carry ripple counter using RS flip-flops and NOR gates.

Figure 5-12 shows two stages of a ripple carry counter implemented with a simple RS flip-flop and NOR gates. When the input is positive, the counter is inactive. In the state shown, both stages start in a logic 0 state. The operation is traced through the first stage (top). In the initial state, Q is 0, $\overline{Q}$ is 1; P, S, and R are 0; and T is 1. When the input goes negative, S goes positive and sets A, which changes $\overline{Q}$ to 0, and Q to 1. This causes T to go to 0. When the input returns to positive, S goes to 0. With both S and $\overline{Q}$ at 0, P goes to 1. This completes a single count cycle. When the next input pulse comes along, the same kind of action occurs, except that the input will generate an output on R because T is now negative. As a result, A changes back to the 0 state.

The second stage (bottom) uses P's output as its input trigger. P cycles once for each two input pulses. Since the binary counter stage changes each time the input goes negative, stage B will change each time P goes negative. This corresponds to A changing from 1 to 0 and is, in effect, the "carry" output.

Figure 5-13 illustrates a modern commercial ripple carry counter, the Signetics S5493/N7493. The S5493/N7493 is a high-speed, monolithic 4-bit binary counter consisting of four master-slave flip-flops which are internally interconnected to provide a divide-by-two counter and a divide-by-eight counter. A gated direct reset line is provided which inhibits the count inputs and simultaneously returns the four flip-flop outputs to a logical 0. As the output from flip-flop A is not internally connected to the succeeding flip-flops, the counter may be operated in two independent modes:

1. When used as a 4-bit ripple-through counter, output A must be externally connected to input B. The input count pulses are applied to input A. Simultaneous divisions of 2, 4, 8, and 16 are performed at the A, B, C, and D outputs as shown in the truth table.
2. When used as a 3-bit ripple-through counter, the input count pulses are applied to input B. Simultaneous frequency divisions of 2, 4, and 8 are available at the B, C, and D outputs. Independent use of flip-flop A is available if the reset function coincides with reset of the 3-bit ripple through counter.

Figure 5-14 illustrates another version of the binary counter—the Fairchild 93191/54191. This is also a 4-bit binary counter, but

COUNT	OUTPUT			
	D	C	B	A
0	0	0	0	0
1	0	0	0	1
2	0	0	1	0
3	0	0	1	1
4	0	1	0	0
5	0	1	0	1
6	0	1	1	0
7	0	1	1	1
8	1	0	0	0

COUNT	OUTPUT			
	D	C	B	A
9	1	0	0	1
10	1	0	1	0
11	1	0	1	1
12	1	1	0	0
13	1	1	0	1
14	1	1	1	0
15	1	1	1	1

(A)

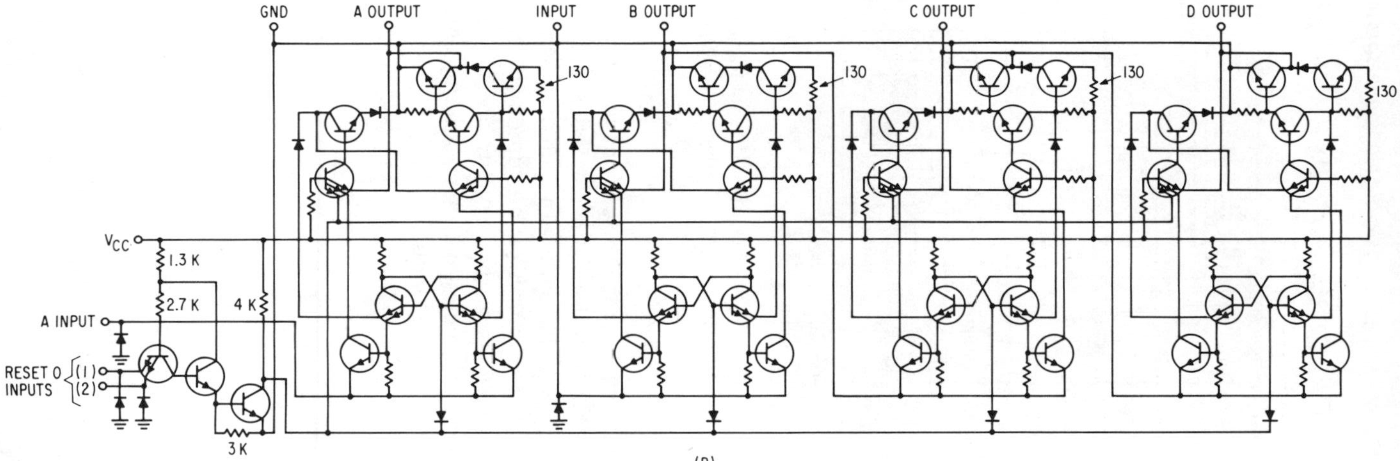

(B)

Fig. 5-13. S5493/N7493 4-bit binary counter. (A) Truth table with output "A" connected to input "B"; to reset all outputs to logical 0, both $R_{0(1)}$ and $R_{0(2)}$ inputs must be at logical 1. (B) Schematic diagram. (Courtesy Signetics Corp.)

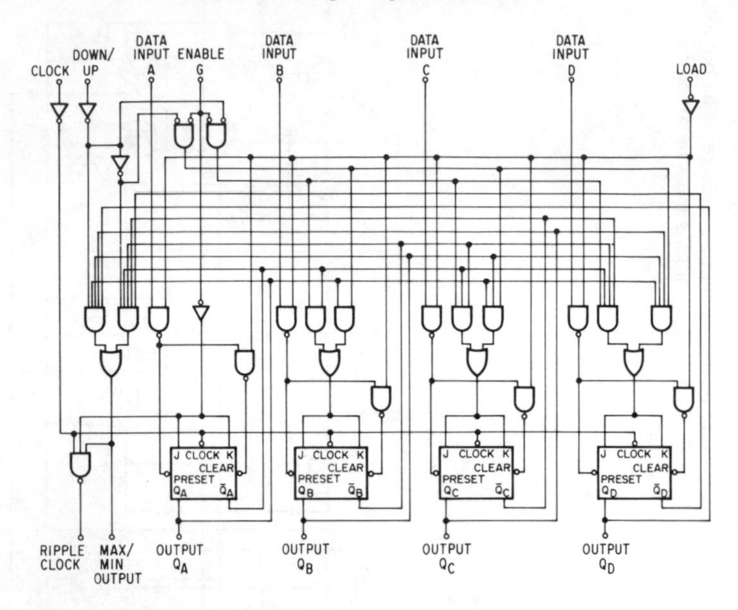

Fig. 5-14. 93191/54191 up-down, 4-bit binary counter. (Courtesy Fairchild Semiconductor)

differs from the previous one by being a "synchronous" type. That is, all the flip-flops are clocked simultaneously so that the outputs change coincident with each other when the proper input conditions appear. Also, this is an "up-down" counter, instead of just an "up" counter, as is the previous one. Thus, depending upon the input signal, the count will increase if triggered with a "low" input, decrease with a "high" input.

6 Troubleshooting

Introduction

This chapter contains information on troubleshooting not only ICs but semiconductor devices as well. Thus, the chapter performs a double task: it can help you in locating faults in ICs and their associated circuits; and it can help you in locating faults in discrete semiconductors and their associated circuits. The material contained in this chapter is based on three premises:

1. You cannot repair an IC; you can only replace it.
2. You are generally familiar with standard schematic diagrams, i.e., those that show individual resistors, capacitors, diodes, tubes, etc.
3. You are not too familiar with logic diagrams although Chapters 2 and 4 have provided a somewhat brief introduction to them.

Premise 1 proscribes troubleshooting within an IC to find out exactly where the trouble lies; it's enough to know that the trouble does lie somewhere in the IC.

Premise 2 permits us to troubleshoot semiconductor (solid-state) circuits down to an IC. Also, if you have the applicable IC data sheet, which shows the actual IC schematic, you can troubleshoot the suspected stage just to make sure it *is* the IC that's at fault.

Premise 3 restricts us to troubleshooting linear circuits and only some simple digital circuits. Also mitigating against troubleshooting complex digital-logic circuits is the equipment required to do such troubleshooting. Such equipment is highly sophisticated and costly. As a minimum you would need an oscilloscope with dual-trace capability; such an instrument runs between $600-$1,000. You would also need a pulse generator, which can run from $150 to $400. Why such complex equipment? Simple: while a good VOM can tell you whether the correct voltages do or do not appear at the pins of an IC, it cannot indicate pulse width, pulse duration, rise time, fall time, etc., all the determining factors as to proper operation of a digital IC.

There are other excellent but less complex equipments you can use for field troubleshooting. The units shown in Figs. 6-2 thru 6-5 are in-circuit testers, are portable, and are *comparatively* inexpensive (none above $300). The Logic Probe (Fig. 6-1) lights up near its tip when the tip is touched to a "high" level, or an open circuit; the light goes out when the tip is touched to a "low" level. The probe also stretches pulses that are 25 ns or wider, to give a light indication of 0.1 second. The light then flashes on or off, depending upon pulse polarity. For static tests, the pulser feeds in a pulse, while the probe indicates whether the appropriate high level or low level exists at the proper IC pins. The Logic Clips shown in Figs. 6-2 and 6-3 clip onto 14- or 16-pin TTL (transistor-transistor logic) or DTL DIP (diode-transistor logic, dual-inline package) ICs and instantly display the logic states of all pins. Each of the Clips' sixteen light-emitting diodes

Fig. 6-1. H-P Model 10525T Logic Probe and Model 10526T Logic Pulser. (Courtesy Hewlett-Packard)

Fig. 6-2. H-P Model 10528A Logic Clip. (Courtesy Hewlett-Packard)

(LEDs) independently follows level changes at its associated pin; a lit LED corresponds to a high logic state. The Logic Comparator shown in Fig. 6-4 also clips onto a 14- or 16-pin DTL or TTL DIP IC. But the comparator operates by connecting the inputs of the test IC and a reference IC (furnished with the comparator) in parallel. Thus, the reference IC is exercised by the identical signals inputted to the test IC. The outputs are compared via the 16 LEDs of the comparator. Here, however, a lit LED corresponds to a difference in levels, indicating a defective test IC.

The tester in Fig. 6-5, on the other hand, can test just about any IC, but is restricted to out-of-circuit use. It is not inexpensive, either, but can be very useful in a shop or laboratory to check out either removed or used ICs, or even new ones before they are used.

This chapter, therefore, is devoted mainly to practical suggestions on troubleshooting solid-state linear circuits, those you would find in a stereo amplifier/receiver, a TV receiver, an AM-FM receiver, or the like. It also includes some simple digital-circuit

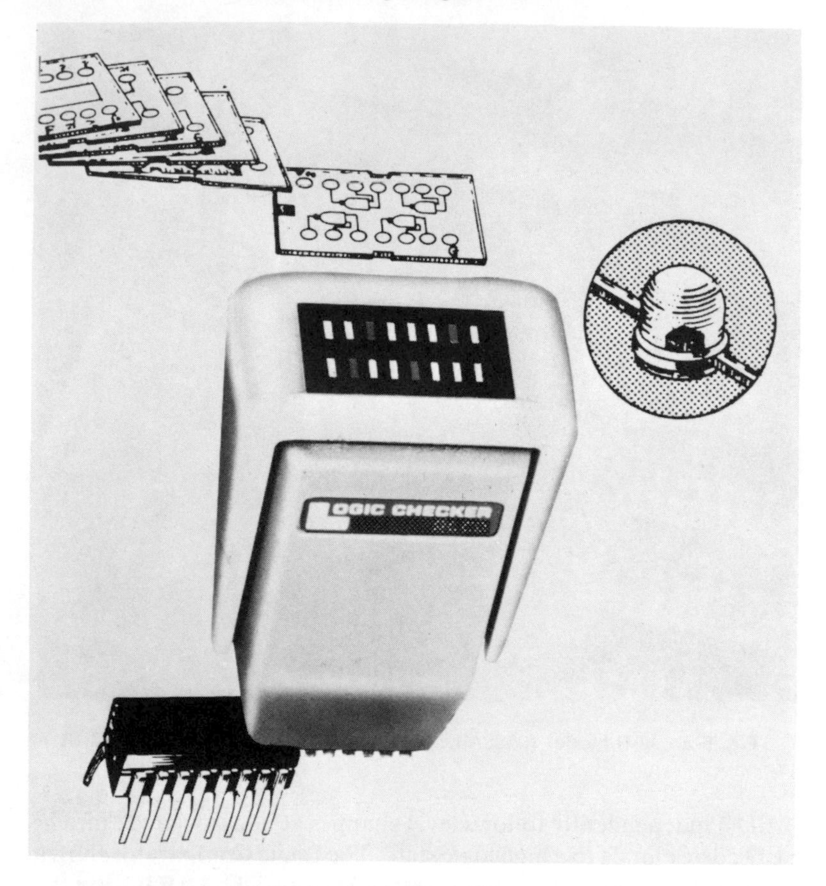

Fig. 6-3. ALCO 201 Logic Checker. (Courtesy ALCO Electronic Products, Inc.)

checks, those that can be performed with a couple of the less-complex equipments described previously.

IC Versus Vacuum-Tube Troubleshooting

The techniques used to troubleshoot ICs and other semiconductors are far different from those used to service vacuum-tube equipment. While the simplest way to troubleshoot vacuum-tube circuits is to replace a suspected tube with a good one, this approach doesn't hold true for semiconductors, discrete or integrated. Unless the IC (or semiconductor) is plugged into a socket, removal is much more difficult than simply pulling a tube: ICs (and

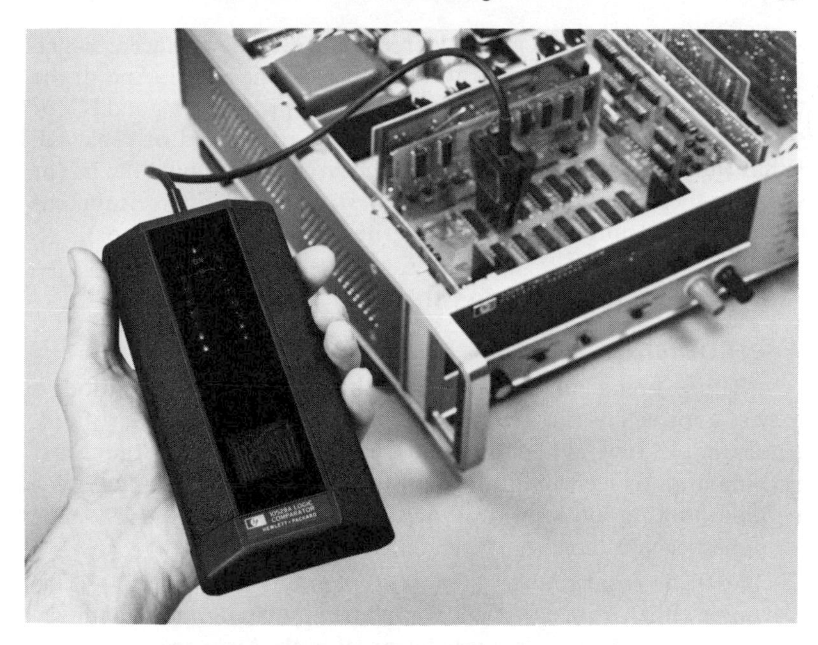

Fig. 6-4. H-P Model 10529A Logic Comparator. (Courtesy Hewlett-Packard)

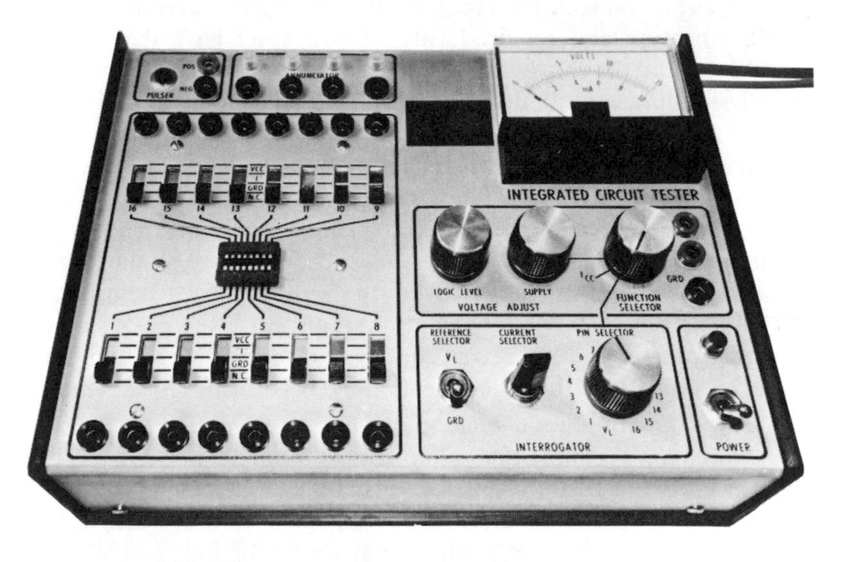

Fig. 6-5. Spectrum Dynamics Model 102 IC Tester. (Courtesy Spectrum Dynamics)

semiconductors) must be unsoldered before being replaced, which may require removal of the entire printed-circuit board. And, if the tools used for desoldering are not employed properly, a good IC (or semiconductor) can be ruined as it is being removed or replaced. What this all comes down to is that the circuit containing the IC (or semiconductor) must be troubleshot *first*, to make fairly certain that the malfunctioning item actually is the IC (or semiconductor).

Basic Troubleshooting Techniques

Prerequisites

Before going any further, one thing must be understood: a logical approach is required in order to isolate and correct *any* fault. For example, it would be foolish to try to repair a TV set without first turning it on to see what the symptoms are. How do you approach a troubleshooting job logically? Actually, the first things to examine are whether you have the "prerequisites for the course."

First, do you know how the equipment operates normally? This can come either from experience or by means of instruction manuals, manufacturer's data, or a combination of both. Also, no matter what experience you have, it cannot possibly encompass every existing circuit, and studies of data sheets for expected waveshapes, signal levels, resistances, voltages, etc., must be made *before* attempting any trouble location and repair. Lastly, if you don't have that much experience, and the data available are sketchy or nonexistent, the best troubleshooting you can do is to realize your shortcomings and refer the job to some agency that can handle it.

Second, do you know what all the controls and adjustments are supposed to do, and how to operate/adjust them? This, too, is based on experience, especially on a piece of simple equipment such as a radio or a black-and-white TV set. And, even for these, you simply do not adjust IF stages unless you have data sheets telling you how much and in what sequence. As for complex equipment, like a good stereo amplifier or a color TV receiver, control and adjustment settings are critical, especially if the components have been in use for a while.

Third, do you have available, and know how to use, the proper test equipment? A simple 1000-ohm/volt voltohmeter (VOM) may be fine for checking vacuum-tube circuits. But for checking transistor biases, which may range anywhere from 0.05V to 0.7V, you really need a transistor checker, such as that shown in Fig. 6-6. Why a special tester? Well, not only does that simple VOM load down the

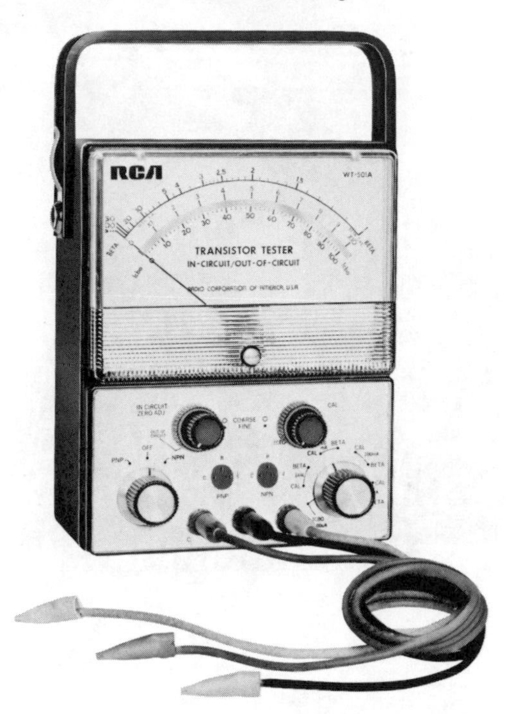

Fig. 6-6. WT-501A In-Circuit/Out-of-Circuit Transistor Tester. (Courtesy RCA Corp.)

circuit so as to give false indications, but its internal batteries may provide enough voltage to burn out the very component being checked. Thus, for troubleshooting any solid-state component, and especially ICs, you need more sophisticated test equipment. Basically, you need a good VOM, such as a Simpson Model 314 Solid-State VOM and/or a Triplett Model 801 Solid-State VOM. The major feature of both instruments, as far as we are concerned, is that they dissipate only *microwatts* (70 μW for the Simpson, 140 μW for the Triplett), making them safe for in- and out-of-circuit IC tests. In addition, both can measure dc voltages as low as 50 mV; for ac ranges the Triplett can go as low as 5 mV, the Simpson to 10 mV.

If you're really serious about troubleshooting, you'll also need a variety of signal sources, i.e., audio generator, RF generator, marker generator, etc., plus the expensive oscilloscope mentioned earlier.

Fourth, do you have the proper tools and know how to use them? Most repairs can be made with the basic tools, such as small

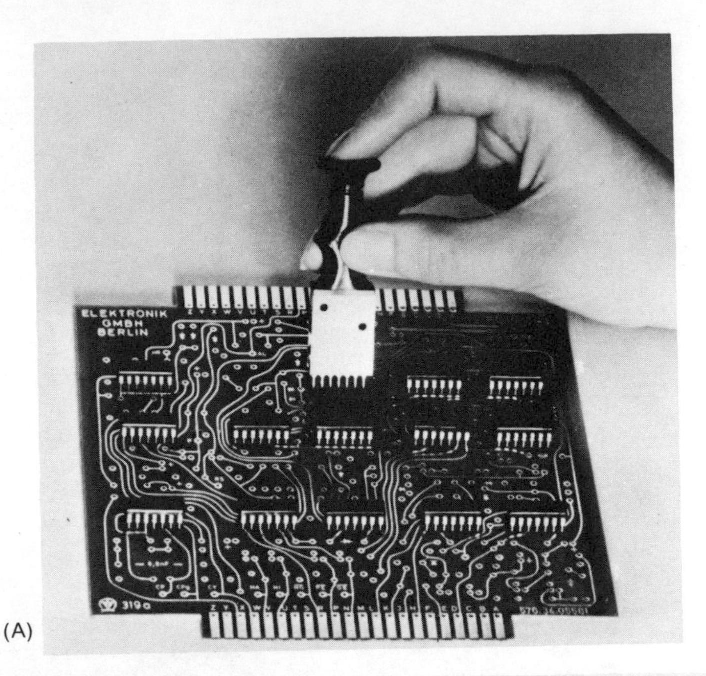

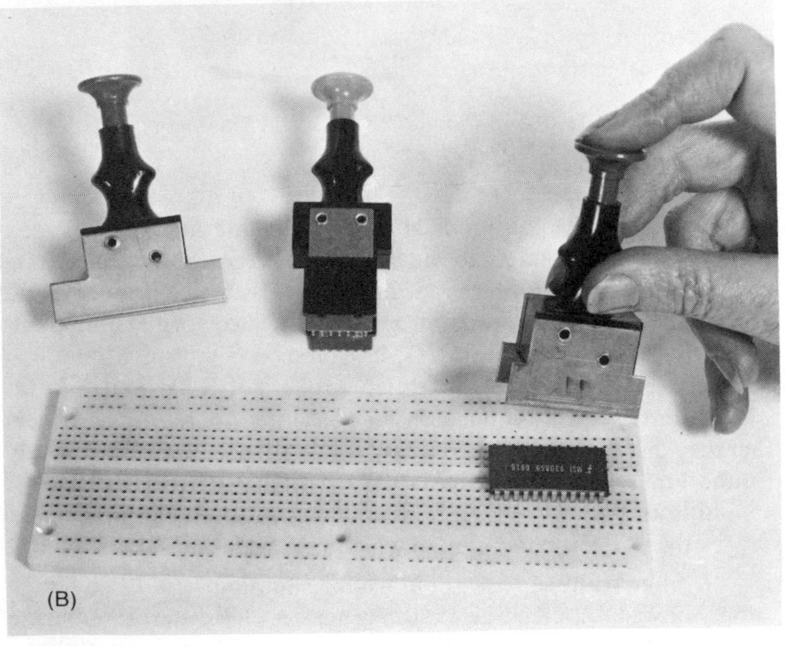

Fig. 6-7. DIP insertion/extraction tools. (A) DIP-A-DIP 14/16-pin IC insertion tool. (B) DIP-A-DIP 24/28 and 36/40-pin IC insertion tools; PUL-A-DIP 24/40-pin IC extraction tool. (Courtesy Micro Electronic Systems, Inc.)

long-nose pliers; small wire cutters, low-wattage soldering iron; screwdrivers; socket wrenches and/or nut drivers; alignment tools; etc. But with ICs (or semiconductors), you will need additional tools, such as desolderers which can remove the solder from many or all IC contacts simultaneously (some IC manufacturers do not recommend desolderers—too much heat applied for too long a time over too great an area); duck-bill long-nose pliers; IC test clips, magnifying glass for checking wiring; DIP insertion/extraction tools (Fig. 6-7); etc.

Lastly, once you have noted the symptoms of the trouble, you must be able to analyze them before you plunge ahead. You must be able to think, and have the patience to sit back and reason out where the trouble most likely lies. In simple terms, don't rush in to replace the speaker just because there's no sound.

Basic Troubleshooting Sequence

There are four steps in the basic troubleshooting sequence:

First, analyze the symptoms. Determine whether you understand all of the symptoms, that is, whether they indicate normal/abnormal/subnormal operation, or complete failure? For example, is the sound from one channel of a stereo receiver louder than the other, or is one channel distorted, or is one channel (or both) completely without sound? Then, unless you have been servicing stereo equipment for a while, check all your available service data; they might contain the very remedy you need. (It also wouldn't hurt to check the most obvious: Is the equipment plugged in and properly hooked up?)

Second, localize the trouble to a specific function or module. For example, if the symptom is "no picture" in a TV set, but there is a raster and sound, you can be fairly sure that the trouble is *not* in the high-voltage, power-supply, sweep, or audio sections, nor in the picture tube itself. In other words, you have localized the trouble to either the IF or video sections. Keep in mind, however, that localization is not always definite: failures in one section can often show up as abnormal symptoms in another section. As an example, a "fuzzy-sounding" speaker is often not the fault of the speaker itself, but of a bad audio-output stage.

Third, localize the specific malfunctioning circuit within the localized function or module. This can be done with the power on via signal tracing or voltage checks; or with the power off via signal injection and/or resistance tests. Signal tracing is performed by examining the inherent signal at a test point with a VOM, oscilloscope, or even a loudspeaker. In tracing, the input probe of the indicating device is moved from point to point, while the signal source remains a constant. Signal injection, on the other hand, involves injecting an artificial signal from some generating device. Here, the indicating device remains fixed at one point, while the signal-injection device is moved from point to point until the faulty circuit is found.

Such testing must be performed with the aid of a schematic diagram which shows waveshapes and/or voltage levels at the points being checked. Otherwise, test results can be meaningless. The trouble may not lie in a signal path. Power supplies neither contain nor propagate signals, hence signal injection and/or tracing are a waste of time. Here, only voltage and resistance measurements are feasible.

Fourth, localize the faulty component. Once you have localized the defective circuit, the first thing to do is to make a visual check of all the components involved. Check for loose connections, cold solder joints, unsoldered joints, cracks in the printed circuit wiring, etc. You can also look for burnt resistors, although in solid-state circuits, with their low voltages (except for power supplies and audio output stages), such occurrences are rare.

Next, check voltages against those called for in the schematic diagram. (If the original symptoms were burning or arcing, this is not feasible. In such cases, the burning/arcing will do your narrowing down for you.) Voltage checks are generally more effective than are resistance checks of solid-state equipment. For example, checking a capacitor that is in parallel with a low-resistance resistor may lead you to think you have a shorted capacitor. Or, if you are using an ohmmeter to check an emitter resistor that has its other end tied (directly or indirectly) to the base, what you might be measuring could be the emitter-base forward resistance since, with the applied voltage of the meter, the transistor could become forward-biased.

Mechanical Considerations

Troubleshooting procedures for solid-state equipment require delicacy, for several reasons. IC, and solid-state circuits in general, are almost invariably printed circuits, so soldering irons must have low-wattage ratings (preferably about 15 W, but no higher than 47 W) to prevent damage. Further, parts in IC circuits are assembled close together, so you need small tools to get at them. Again, although semiconductors and ICs are inherently reliable, due to their generally low-power requirements, some are not too rugged physically. Leads are usually fragile and can be snapped easily while being soldered or unsoldered. You might even damage the IC itself if you pull the lead or pin the wrong way or vibrate either. Lastly, because of their low-operating voltages, small bias voltages can cause IC burnout. So be careful to avoid inadvertent short circuits while troubleshooting with the power on.

A good percentage of the problems in repair work, and even a few in original factory assembly, can be traced to poor soldering. Unless you are thoroughly experienced in soldering, it might be helpful to note the following points before starting to make any repairs: Always use *resin core* solder, such as Ersin Multicore 60/40 or Kester No. 44 (0.032 in. diameter). And, always use a *clean*, well-tinned iron, *whose tip is grounded.*

A *simple* mechanical connection is required before soldering. A single bend of the wire, squeezed tightly to its terminal point, will do. On the other hand, with printed circuits just a light touch of solder will often suffice (too heavy, and you might create short circuits between the printed wiring).

P-c paths are highly sensitive to heat. For example, too much heat applied for too long a time will cause the printed circuit to pull away from the board when you pull the iron away. As for the components, use only enough heat to melt the solder. A device like a paper clip or mini-alligator clip mounted on the lead between the device and its terminal point will help drain away the heat. Or, you can just lay the metal shaft of a small screwdriver on the lead. Whatever you use, do not remove it until the solder has solidified on the joint. Lastly, for DIP ICs, the foregoing will not always hold—the connection pins may be too short to heat sink. So, if you have the room, put in an IC socket instead. You can then plug in the IC just as you would a vacuum tube (see Fig. 1-7). If you do not have the room, be careful with the iron, or the desoldering tool (it can also be used to resolder if you melt the solder first, then put the IC in).

If a component seems loose on a p-c board, you do not have to pull out the board to effect your repair. To fix looseness, simply pull the loose lead up tight against the foil side of the board, then apply the iron to the lead. Let go when you feel the lead starting to come out of the board.

If a component is to be removed and replaced with a new part, cut the leads near the body of the component. This will free the leads for individual unsoldering, or, if you have the room, for simply soldering the new component to the old leads. Otherwise, grip the cut lead with duck-bill pliers, then apply the tip of the iron to the connection at the back of the board and pull the lead out gently. If you can't or don't want to remove the board to get at the back, apply the iron to the cut part of the lead and push it out through the back when the solder has melted sufficiently being careful not to take the p-c wiring with you as you push.

If the component is to be removed for test and possibly replaced, do not cut the leads. Instead, grip the lead from the front with the pliers and apply the iron to the connection at the back of the board. Then lift the lead straight out.

A note of caution here. The two removal procedures just discussed assume that the leads went straight into the p-c board, and were not bent over at the back of the board. It would be wise to examine the joint for this condition *before* you start using the iron. If the lead *is* bent, apply the iron to the joint and, when the solder has melted, gently pry the bend away from the board with a pick or X-acto knife. Straighten the bend as much as possible so that it will come out of the board easily. When the lead comes out it should leave a clean hole. If not, clean the hole by reheating the solder and using a needle, toothpick, or enameled wire to ream out the hole.

If the component you are replacing is a small transistor, use a socket wrench, or nut starter, as a holder to guide the transistor into place. Most transistors will fit into 3/16 in., 7/32 in., or 3/8 in. sockets. Do not jam the transistors into the sockets—you may never get them out again, you may dent the casing, or you may break off a few leads while you are pulling the socket off the transistor. When replacing a MOSFET or an IC, make sure that the hand you use to hold the device is *grounded*, e.g., through a metallic wristband.

If you are trying to remove excess solder from a joint without a desoldering tool, take a piece of copper braiding like that found in

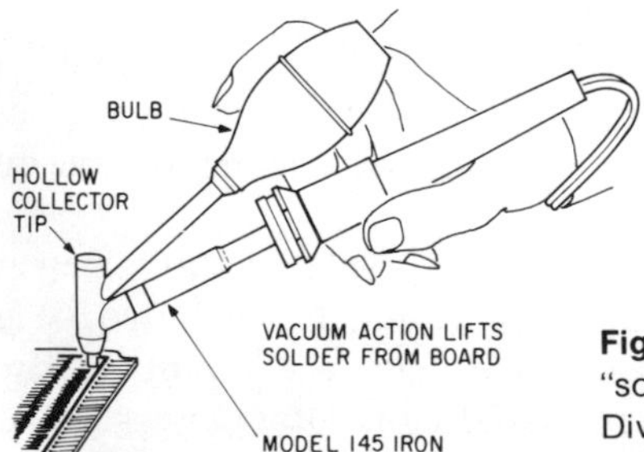

Fig. 6-8. Using the Ungar No. 7825 "solder gobbler." (Courtesy Ungar Div. Eldon Industries)

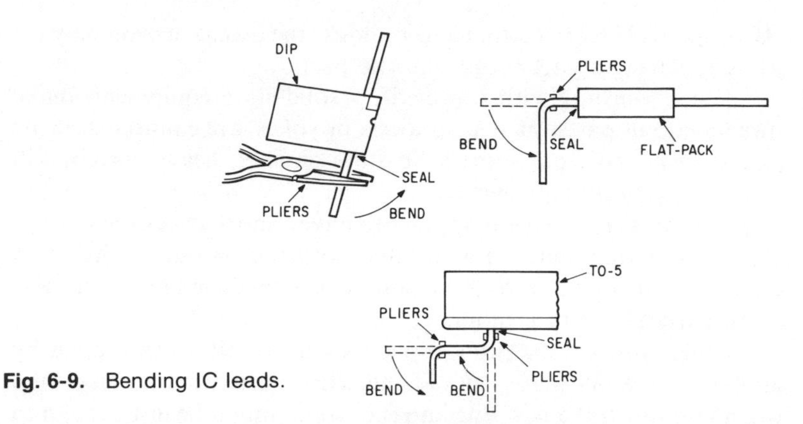

Fig. 6-9. Bending IC leads.

coax shielding, dip it into some solder paste, and put it between the iron and the connection. The braid will then sop up the excess solder like a sponge. An alternative method is to use a "solder gobbler," like that shown in Fig. 6-8. The gobbler consists of a soldering tool, collector tip, and bulb. First the bulb is squeezed, then the collector tip is placed on the connection, and when the solder melts, the bulb is released, thereby sucking up the solder into the collector tip; you can force the solder from the tip by squeezing the bulb once more. The gobbler is especially effective for removing ICs from a circuit.

When replacing an IC, the mounting pattern of the original IC must be followed *exactly*, as there is rarely enough space in IC equipment to do otherwise. You may, therefore, find it necessary to bend the leads of the replacement IC. If so, it is very important that the leads be supported and clamped between the bend and the seal at the IC case. Otherwise, you may break the seal or damage the lead plating. Long-nose or duck-bill pliers can be used to hold the lead as shown in Fig. 6-9. In no case should the radius of the bend be less than the lead diameter or, in the case of rectangular leads, less than the thickness of the lead. It is also important that the ends of the bent leads be perfectly straight and parallel to assure proper insertion into the holes in the p-c board.

Miscellaneous Suggestions

If you are making in-circuit repairs, *pull the plug.* Many of today's solid-state circuits are the "instant-on" type, meaning that some circuits are "live" even though the power switch has been turned off. And, aside from getting a nice shock, the transients you may get while replacing a part could ruin the part.

When working with powered-up solid-state equipment, make sure all circuit parts, such as speakers or yokes, are connected. If the load is removed from some solid-state circuits, heavy current will flow, resulting in damaged ICs.

If a transistor element appears to have a short, check the settings of any operating controls associated with the particular circuit. A volume control set to zero, for example, can give the same reading as a short from base to ground.

When you are checking capacitors in-circuit, do not do it by jumping a new one across the old one. The transient surge that occurs when you touch the new one into the circuit might be just enough to damage or ruin an IC (or semiconductor) in that circuit.

Some additional hints on capacitors in IC circuits. If you have a failed IC, check to see whether the circuit is using capacitors for input coupling to the IC, output coupling from the IC, for noninverting feedback, or for output-line loading. Capacitors, especially if they are low-impedance types, have a habit of presenting instantaneous current levels at the input/output of a device that are greater than the rated dc levels. While a good IC should handle these peaks for a while, the peaks cause premature aging and early breakdown. The really bad feature is that the inclusion of capacitors in this manner is a design error or cost-cutting procedure, so all you can do is replace ICs, or figure out a better design for that circuit.

Check to see whether a failed IC is driving some sort of coil or incandescent lamp. Here, again, the designer may have decided to cut corners by using the IC's maximum rated output current as his driving source. This also ages the IC prematurely, again necessitating frequent replacement or your own redesign.

Lastly, where there is no obvious cause for a low voltage at some point in the circuit, or an abnormally high resistance exists, take a magnifying glass and look for cold solder joints, p-c wiring breaks or hairline cracks, or even tiny gobs of solder that might be shorting a couple of closely-run p-c wires. Breaks or cracks can be repaired by the judicious use of a little solder, or even by a discrete piece of wire, if the space is available.

Solid-State Troubleshooting

Aside from ICs, which contain a multiplicity of circuits, the two major semiconductor devices are diodes and transistors. (In monolithic ICs, diodes are often formed by laying down a transistor pattern wherein the collector and base are shorted together to form the cathode of the diode, and the emitter forming the anode.) Since it is simpler to test diodes, we will begin with them. But whether you are testing a discrete or an IC diode, the procedures are the same.

Solid-State Diodes

Diodes have three basic requirements: (1) they must be able to pass current in one direction (forward current) while preventing or limiting current flow in the opposite direction (reverse current); (2) for a given forward current the voltage drop across the diode should not exceed a specific value; and (3) for a specific reverse voltage, the reverse current should not exceed a given value.

All of the foregoing are best determined out of circuit, using an oscilloscope and appropriate data sheets, or a diode tester. However, an ohmmeter can be used to good effect in checking a diode's current-passing/limiting capabilities. A good diode will exhibit high resistance in the reverse direction, and low resistance in the forward direction.

If the resistance in the reverse direction is low, the diode is probably leaking. If low in both directions, the diode is most likely shorted. If high resistance is found in both directions, the diode is probably open. The basic factor in determining a good diode, in addition to high-reverse/low-forward resistance, is the *ratio* between the two resistances. The actual ratio depends upon the type of diode; for small, signal diodes the ratio should be at least 1:100, forward-to-reverse (front-to-back is the more common name). For power diodes, a satisfactory ratio is at least 1:10. For IC diodes, the ratio should be at least 1:30. Incidentally, to check IC diodes you *must* have a schematic of the IC itself, since most equipment schematics show an IC simply as a sideways triangle.

Transistors

The two most common types of transistors, be they discrete or integrated, are the bipolar (base, emitter, and collector) and the unipolar or field-effect (gate, source, and drain) types. We will begin with the bipolars.

Bipolar Transistors Basically, linear circuits consist of resistors, capacitors, etc., and bipolar transistors, as already stated. The following covers how to track down troubles to these transistors, and how to then check the transistors themselves. Again, remember that

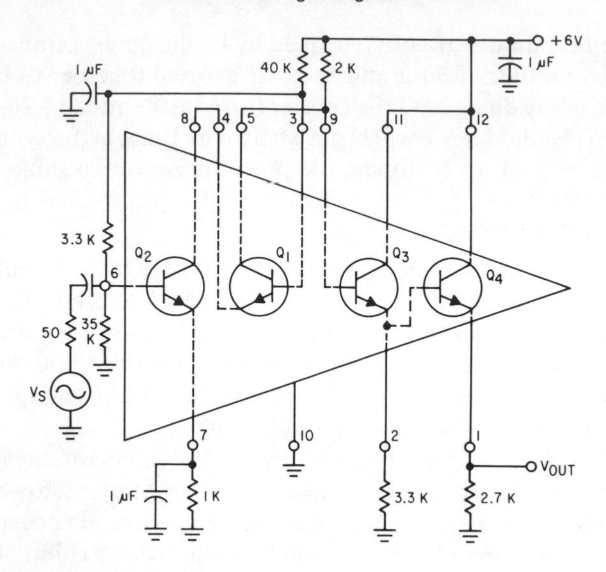

Fig. 6-10. Schematic diagram for a CA3018 Cascode Video Amplifier (dotted lines inserted by author for use in this text only). (Courtesy RCA Corp.)

on a schematic diagram the symbol for an IC is usually a sideways triangle, as shown in Fig. 6-10. The dotted lines have been inserted by the author, and are *not* normally given in the schematic. So, as with IC diodes, you must have the IC schematic if you want to check out an IC transistor.

Bipolar transistor circuits are best tested via voltage checks, using the applicable schematic diagram, and a sensitive VOM, such as a Simpson Model 314 or a Triplett Model 801.

Figure 6-11 shows the basic connections for both p-n-p and n-p-n transistor circuits; the coupling and bypass capacitors have been omitted for simplicity. In practically all transistor circuits, the emitter-base junction must be forward-biased to obtain current flow through a transistor. In a p-n-p, this means that the base must be made more negative than the emitter. The emitter-base junction will then draw current, causing heavy electron flow from collector to emitter. In an n-p-n, the base must be made more positive than the emitter for emitter-to-collector current flow.

The most common way to measure transistor voltages is between ground and the element; manufacturers' data generally specify transistor voltages this way. For example, all the voltages for the p-n-

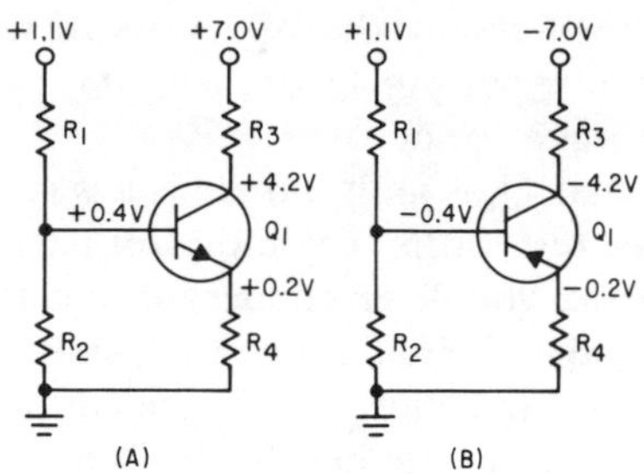

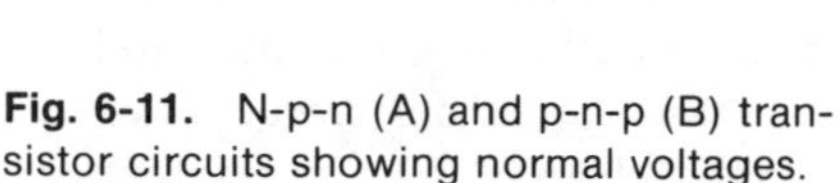

Fig. 6-11. N-p-n (A) and p-n-p (B) transistor circuits showing normal voltages.

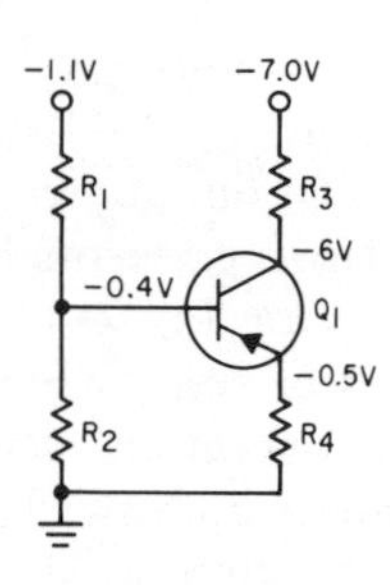

Fig. 6-12. P-n-p transistor circuit of Fig. 6-11 with abnormal voltages.

p of Fig. 6-11 are negative with respect to ground. The following demonstrates how voltages measured at the transistor elements can be used to analyze failure.

Assume that the p-n-p circuit of Fig. 6-11 is measured and the voltages found are those of Fig. 6-12. The first clue that something is wrong is that the collector voltage is almost the same as the source voltage at R_3, indicating that very little current is flowing through R_3. The resistor could be defective, but the trouble is more likely caused by a bias problem. The emitter voltage depends mostly upon the current flowing through R_4; thus, unless the value of R_4 has changed drastically, the problem is one of base bias.

The next step, then, is to measure the voltage at R_1. If the bias source voltage is, say, -0.9V instead of the required -1.1V, the problem is obvious; there is probably a defect in the power supply. If the bias source voltage is correct, then either R_1, R_2, or Q_1 is defective.

The next logical step is to power-down the equipment and measure R_1 and R_2. If either is incorrect, there's your solution. If both values are correct, check R_4, just to be sure. However, it is more likely

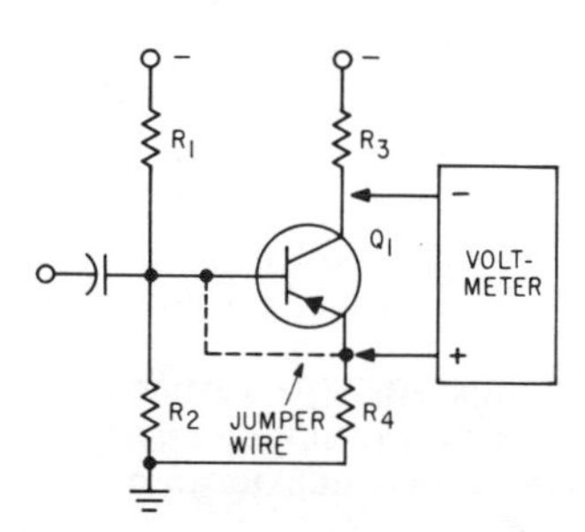

Fig. 6-13. In-circuit transistor test setup.

that Q_1 is faulty. (A short note on the foregoing: the capacitors are not shown in Figs. 6-11 and 6-12. So, before you pull Q_1 or the IC it is part of, check out any capacitors that might be across R_1–R_4.)

Having reasonably assumed that Q_1 is faulty, there is a way to make sure: Fig. 6-13 illustrates the test setup, if you do not have a transistor tester (Fig. 6-6). With the voltmeter connected and the circuit operating, measure and note the difference potential, emitter-to-collector. Now shut off the power and short-circuit the emitter-base junction, as shown by the dotted-line jumper. Reapply power, and remeasure the emitter-collector potential: if it is not much higher than the original measurement, the transistor is defective.

Transistors (discrete) can also be checked out of circuit via simple resistance measurements (again assuming you do not have a transistor checker, Fig. 6-6). From the semiconductor theory of Chapter 3, we know that bipolar transistors are either p-n-p or n-p-n types. This means that there are two p-n junctions in each type. And, since a p-n junction is, in effect, a semiconductor diode, it can be tested in the same way. So, to check an n-p-n type, the ohmmeter leads are first connected as shown in Fig. 6-14A, then with the leads reversed so that the meter COM lead is connected to the base and the + lead to the emitter. The ratio of front-to-back resistance should be at least 1:30. If this step checks out, the process is then repeated for Fig. 6-14B and Fig. 6-14C, each time reversing the leads to determine the front-to-back ratios.

As for p-n-p types, the procedure is exactly the reverse. That is, if it were a p-n-p in Fig. 6-14A, the meter COM lead would be connected to the base and the + lead to the emitter, and then reversed.

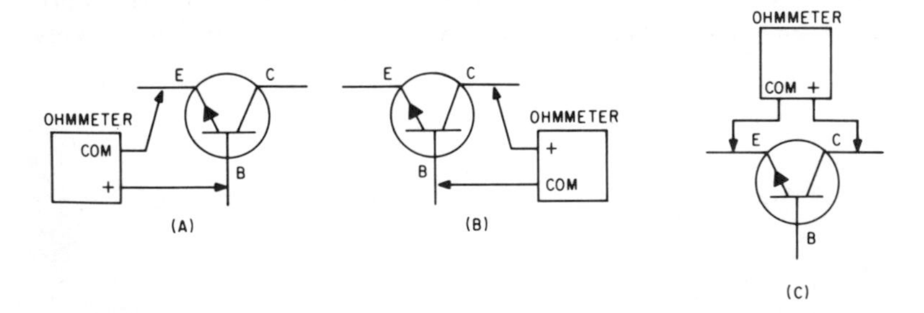

Fig. 6-14. Resistance testing an n-p-n transistor. (A) Testing Emitter—base forward resistance. (B) Testing Collector—base forward resistance. (C) Testing Collector—emitter forward resistance.

Unipolar (Field-Effect) Transistors Unipolar (field-effect) transistors can be checked in a variety of ways. A good VOM will tell you whether the proper source and output voltages exist. Or, you can use an ohmmeter to check front-to-back ratios. But, with IC field-effect transistors (FETs) you must be extra careful. Their performance depends on the relative perfection of the insulating layer, if any (some IC FETs do not have an insulating layer) between the gate (equivalent to the base of a bipolar transistor) and the active channel. Should this layer become punctured by inadvertent application of excess voltage to the external gate connection, like the battery voltage across the leads of an ordinary 1000-ohm/volt VOM, the damage done is irreversible.

If the FET is part of an IC, you can also use the tester shown in Fig. 6-5. This unit can perform static tests, both dc and functional, on flatpacks, TO-5 and TO-8 types, and up to 16-pin DIP types. It is designed for use by engineers, technicians, and servicemen, hence may fall within your capacity. And, since many equipments are designed with plug-in ICs, it could come in very handy.

Finally, the circuit shown in Fig. 6-15 will enable you to determine whether your IC MOSFET is open or shorted. It is a single go/no-go circuit that will test out-of-circuit, n-channel depletion types or p-channel enhancement types (see Fig. 3-16 for symbols for both). The substrate (B) and source (equivalent to the emitter of a bipolar) are connected to terminal 1, the gate to terminal 2, and the drain (equivalent to the collector of a bipolar) to terminal 3. If the MOSFET is a dual-gate type, test each separately.

For n-channel depletion types, if the lamp lights with the switch open, then goes off when the switch is closed, the transistor is "good," i.e., neither shorted nor open. If the lamp lights whether the switch is

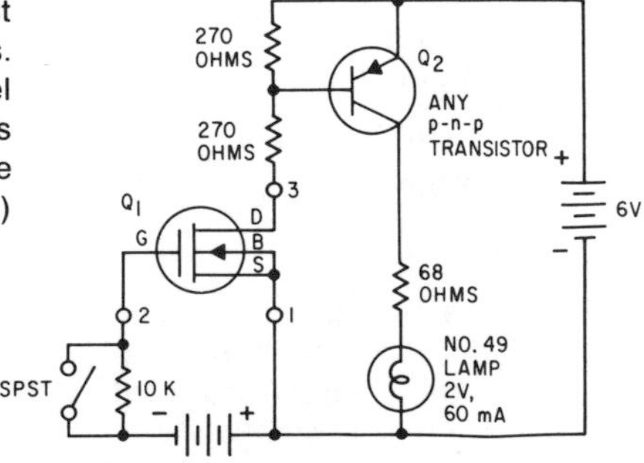

Fig. 6-15. Simple go/no-go test circuit for MOS transistors. (*Note:* Q_1 is an n-channel depletion type shown in this example. p-Channel would have the arrowhead pointing outward.) (Courtesy RCA Corp.)

open or closed, the transistor is shorted. If the lamp does not light at all, regardless of switch position, the transistor is open. For p-channel enhancement types, the lamp should light with the switch closed, and go out when the switch is opened. Otherwise, the test is the same.

Digital (Pulse) Circuits

As with linear circuits, digital circuits also consist of resistors, capacitors, diodes, and transistors. However, here the similarity ends: digital logic diagrams do not show discrete transistors, they show symbols such as those of the logic ICs of Chapters 2 and 4. These symbols represent *functions* (AND, OR, etc.), with the functions being performed by combinations of resistors, diodes, and transistors, as shown in Figs. 6-16 and 6-17.

Checking complete digital circuits is far beyond the scope of this book. As already noted in the beginning of the chapter, to really check a digital circuit you must be very familiar with *all* the complex logic symbols and how they interrelate in a logic diagram, and you must have the proper test equipment. Testing, therefore, is limited to checking a few of the basic circuits shown in Figs. 6-16 and 6-17 using the Logic Probe and Logic Pulser shown in Fig. 6-1.

Testing an AND Gate Assume the AND gate shown in Fig. 6-16A, connected in a DTL or TTL system.

1. Connect the probe to a source of +5V dc; the band should light up.
2. Place the probe tip on any gate input pin. If a "high" level is expected at the pin, the band should remain lit; if a "low" level is expected, the light should extinguish. If a series of pulses is expected, such as clock pulses, the band should be lit dimly if the expected frequency is 1 MHz or less, and lit dimly or go out momentarily if the frequency is above 1 MHz (up to 20 MHz).
3. If the expected levels or pulse trains appear at the proper input pins, shift the probe to the gate's output pin. Depending upon the logic system used, the band should remain lit for a high ("1"—positive logic, "0"—negative logic), and extinguish for a low (vice versa).

Fig. 6-16. Basic logic gates in circuit form (*right*) and with logic symbol (*left*). (A) Three-input AND gate. (B) Three-input OR gate. (C) Three-input NAND gate. (D) Three-input NOR gate. (E) Two-input Exclusive-OR gate.

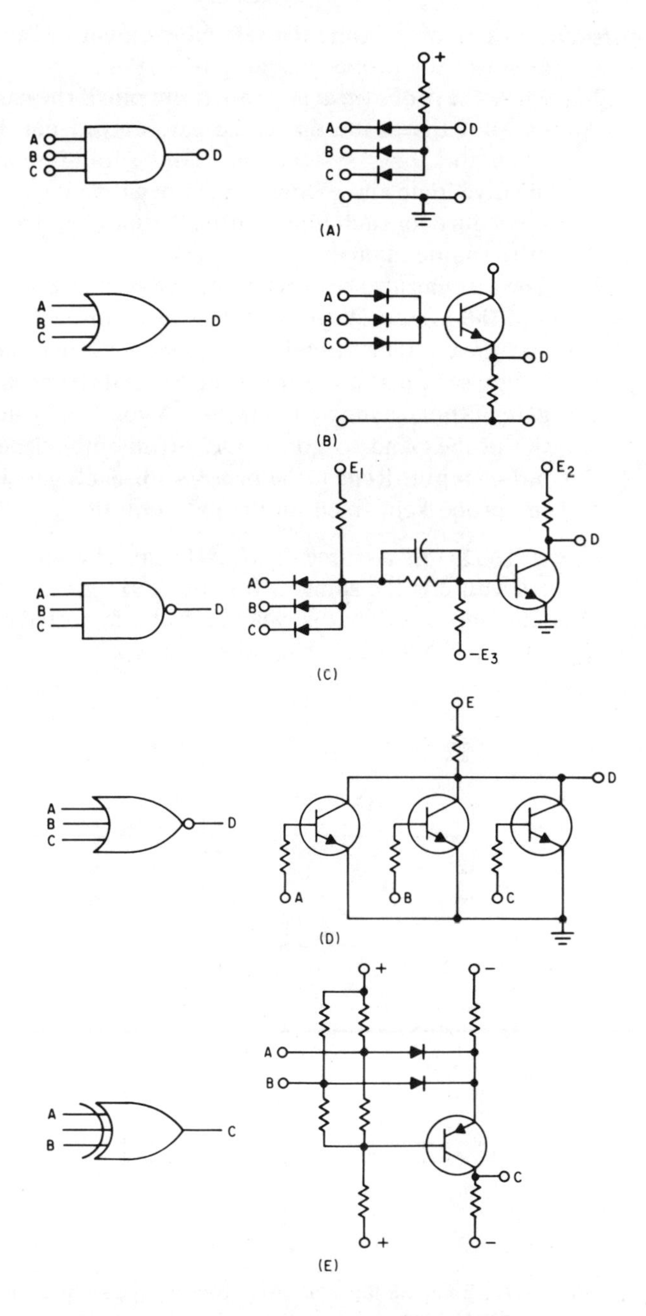

A
B
C
D
+
(A)
(B)
E1
E2
-E3
(C)
E
(D)
-
(E)

Testing an OR Gate Assume the OR gate shown in Fig. 6-16B.

1. Connect the probe to source of +5V dc.
2. Place the probe tip at any gate input pin; if the band remains lit, shift the probe tip to the gate output pin. Depending upon the logic system, the band should remain lit for positive logic and extinguish for negative logic. Repeat the procedure for each input pin, each time checking the output in the same manner.
3. The gate can also be checked statically, using both the probe and the pulser. With the pulser also connected to +5V dc, and the equipment under test powered down, simply place the pulser tip at any gate input pin and the probe tip at the gate output, as shown in Fig. 6-1. A good OR gate will cause the probe band to go on and off in coincidence with the pulser input. Repeat the process for each gate input, with the probe kept fixed on the gate output.

Testing a NAND Gate Assume the NAND gate shown in Fig. 6-16C. The test procedure is the same as for the AND gate, but the result should be the opposite. That is, where a "1" is expected at the AND gate output, the NAND gate should produce a "0."

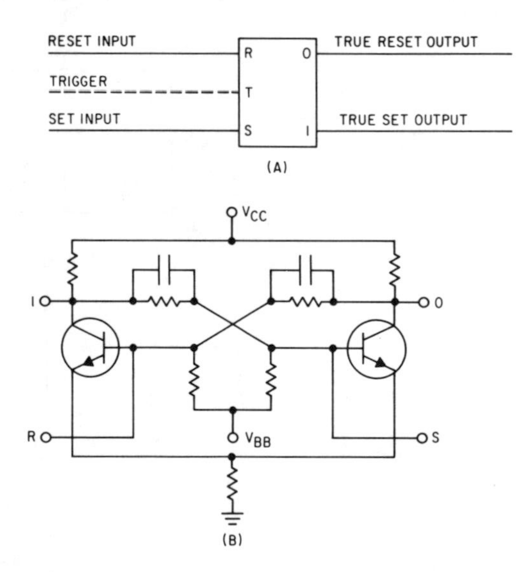

Fig. 6-17. Basic RS flip-flop. (A) Flip-flop logic symbol. (B) Flip-flop in circuit form.

Testing a NOR Gate Assume the NOR gate shown in Fig. 6-16D. The test procedure is the same as for the OR gate, but the results should be the opposite. That is, where the OR gate should cause the probe band at the output to go on then off, the NOR gate should cause the band to go off then on.

Testing an Exclusive-OR Gate Assume the Exclusive-OR gate shown in Fig. 6-16E. The procedure and results are the same as in the third step of the OR gate test.

Testing a Flip-Flop Assume the simple RS (reset-set) flip-flop shown in Fig. 6-17.

1. Connect the Logic Probe and Logic Pulser to a source of +5V dc.
2. Place the probe tip at the *set output* of the flip-flop, then place the pulser tip at the *set input*. The probe band should remain lit.
3. With the probe tip still at the *set* output, move the pulser tip to the *reset input*. The probe band should now extinguish. You can also check the *reset* function the same way, with the probe tip placed at the *reset output*.

Glossary

Integrated Circuit Terminology

Active Element Component which acts upon (amplifies, rectifies, switches, etc.) an applied signal so as to change the signal's basic characteristics. Examples are transistors and diodes.

Batch Processing Manufacturing method whereby a particular process operates on a large number of components simultaneously.

Bipolar Transistor Transistor whose operation is based on *two* types of charge carriers: holes and electrons.

Bucket-Brigade Device See *Charge-Coupled Device.*

Buried Layer A layer of n+-type or p+-type impurity diffused into the semiconductor substrate *beneath* the epitaxial layer. + denotes a heavier-than-normal concentration.

Cermet Abbreviation for *Cer*amic-*Met*al. A strong alloy of a heat-resistant compound, such as titanium carbide, and a metal, such as nickel.

Channel An area beneath the silicon dioxide layer of a planar transistor that has been switched from one type of conductivity to the opposite type by the action of dopants hitting the dioxide layer.

Charge-Coupled Device IC wherein electrical conduction between the circuit components is performed by charge transfer rather than via the metalization.

Chip A tiny piece of semiconductor material, broken from a semiconductor wafer, on which one or more integrated electronic components are formed.

Compatible IC A monolithic IC made with a thin-film layer of passive components on top of the semiconductor IC.

Crossover In a schematic diagram, that point where two or more lines cross another line, but are not electrically connected, i.e., no connection "dot" is shown. In an IC, it is the metalization that does physically what the schematic crossover does on paper.

Deposition The process of depositing one material on top of another.

Die See *Chip*.

Dielectric Isolation A process whereby individual IC components are separated by dielectric isolating layers, usually silicon dioxide, instead of the usual p-n junctions.

Diffusion Process A process used to spread dopants (impurities) throughout a semiconductor material.

Digital IC An IC used for logic-circuit (computer) operations.

DIP Dual-Inline-Package. A rectangular IC container where the leads form two rows along the long sides of the rectangle, like a centipede's legs.

Discrete Components that are fabricated and/or packaged separately (e.g., standard resistors, capacitors, diodes, transistors, etc.).

Dopant An impurity diffused into a semiconductor material to form the required n- or p-type conductivity regions.

Double Diffusion A process used to separate the individual layers of a monolithic, n-type structure. p-Diffused isolation zones form p-n junctions, which then act as reverse-biased diodes.

Epitaxial Growth Deposition of single-crystal n- or p-type material, or intrinsic silicon, on top of a substrate.

Epitaxial Layer The crystal layer deposited or "grown" via epitaxial growth (deposition).

Evaporation A process whereby metalization or thin-film components are vaporized then deposited through a photomask onto a substrate.

FET Field-Effect Transistor. Transistor that uses either semiconductor holes or electrons, not both, as the charge carriers. Also known as Unipolar Transistor.

Flatpack IC container that is a flat square, with the connection leads coming out of two opposite sides of the square. Thinner than a DIP.

Forward Bias Forward bias exists when, with battery + connected to n-type material and battery – connected to the p-type material of a p-n junction, electrons flow from battery + through the n-type, through the junction, through the p-type, and back to battery –. This state will exist so long as the battery remains connected.

Hometaxial Abbreviation for Homogeneous Axial. A single-diffused transistor with a base region of homogeneous resistivity from emitter to collector.

Hybrid (1) ICs made by two different processes (e.g.,

semiconductor and thin film), or IC plus discrete. (2) Multi-chip circuits within a single IC container (package).

IGFET Insulated-Gate Field-Effect Transistor. See MOSFET.

Impurity See *Dopant*.

Ion Implantation A process whereby ions are implanted into an IC during manufacture to change the IC characteristics (e.g., from depletion to enhancement mode).

Isolation Diffusion See *Double Diffusion*.

JFET Junction Field-Effect Transistor. Unipolar transistor that uses a reverse-biased p-n junction as the control electrode (gate).

Linear IC Any IC device that functions in conjunction with analog signals rather than digital signals (e.g., an amplifier).

LSI Large-Scale Integration. Integration on a single chip of 100 or more circuits equivalent in complexity to a logic gate.

Mesa Transistor A transistor whose epitaxial layer is shaped like a *mesa*, i.e., a flat-topped mountain.

Metalization The process whereby metal is evaporated onto an IC to form the component interconnections and contacts to the outside world.

Monolithic Literally, "a single stone." In ICs, a device in which all circuit elements are constructed and interconnected on or within a single piece of silicon.

MOS Metal-Oxide Semiconductor. A semiconductor with a layer of oxide on its top surface and a metal on top of the oxide. See also *MOSFET*.

MOSFET Metal-Oxide Semiconductor Field-Effect Transistor. A field-effect transistor where the channel is the semiconductor, the oxide layer is an insulator between the semiconductor and the gate, and the gate is a metal electrode deposited on the oxide layer and functions to control the action of holes/electrons in the channel (between source and drain).

MSI Medium-Scale Integration. Integration of from 10 to 100 logic gates on a single chip.

MTOS Metal-Thick Oxide Semiconductor. MOS transistor with a thicker than usual oxide insulating layer between the gate and the active channel to prevent gate damage due to overvoltage or improper handling during repair or replacement.

NMOS n-type MOS.

Nonlinear IC See *Digital IC*.

n-Type Material Semiconductor material to which impurities have been added so as to obtain excess free electrons.

OIC Optoelectronic Integrated Circuit. An IC whose electronic circuits produce an output that is visible.

fall from 90 percent of its maximum value to 10 percent after the input signal has been cut off; the time elapsing while the output voltage falls to within 10 percent of the steady-state value.

Fan-In The total number of inputs to a particular circuit.

Fan-Out The total number of load-connected outputs of a particular circuit; or the total number of loads connected to the single output of a circuit.

Flip-Flop (F/F) A circuit which can only be in either of two stable states: conducting or nonconducting ("on" or "off"). A flip-flop normally has two inputs—Set and Reset (clear). Either will change the state, which will then remain constant until another Set/Reset input is applied. However, if simultaneous Sets or Resets are applied, the flip-flop will go into an indeterminate state. Also known as Bistable Multivibrator, as opposed to Astable, which is simply a self-excited oscillator that generates complementary outputs and is always "on."

Four-Quadrant Multiplier Device or circuit capable of multiplying two variables, X and Y, in all four possible combinations: (X) (Y), (X) (-Y), (-X) (Y), and (-X) (-Y).

Gate Logic Switch, i.e., acts as a short or open circuit, depending upon the input signal(s).

Gating On-off switching.

Glitch Unwanted transient (e.g., a "spike" on input or output square wave).

Handshake Within a communication system, the acknowledgment that all parties are "on the line" and that communication can now proceed; e.g., when you dial someone and he answers "Hello," that dial/"Hello" process is the "handshake."

High (1) Voltage input of sufficient level to operate a gate, i.e., a logic-1 input. (2) Voltage output representing a logic-1 in a positive-logic system, and a logic-0 in a negative-logic system.

HNIL High-Noise-Immunity Logic. Logic circuits that are not disturbed by high-electrical-noise environments.

HTL High-Threshold Logic. See *HNIL*.

Inclusive-OR Gate See *OR Gate*.

Inverter A circuit whose output is of opposite polarity to its input. Also known as the Common-Emitter configuration.

J-K Flip-Flop Flip-flop whose outputs are fed back to the inputs to prevent one of two identical inputs from setting the flip-flop into an indeterminate state, neither "on" nor "off." See *Flip-Flop*.

Latch Circuit Usually, a feedback loop used to maintain the existing state of a symmetrical digital circuit.

Location A register in a Main-Frame Memory.

Logic Circuit (1) Computer switching circuit. (2) Circuit that operates only in response to logic-1s or logic-0s, i.e., pulses or square waves.

Logic-1 In a positive-logic system, a logic-1 is a signal at a sufficiently high voltage level to turn the system "on." In a negative-logic system, a logic-1 turns the system "off."

Logic-0 Reverse of Logic-1.

Low Reverse of High.

LPTTL Low-Power Transistor-Transistor Logic. TTL that dissipates very little power, i.e., draws very little current. See *TTL.*

LSA Large-Scale Array. IC containing a large number of independent IC components that can be employed as the user sees fit.

Main-Frame Memory The primary data storage section of a general-purpose computer. Also called Main Memory.

MECL Motorola ECL.

Memory A device that can store an electrical pulse for later use.

NAND Gate NOT-AND Gate. Reverse of an AND Gate.

Negative Logic A logic system whereby a logic-0 input energizes the circuits; i.e., logic-0 is high, logic-1 is low.

NOR Gate NOT-OR Gate. The reverse of an OR Gate.

OEM Original Equipment Manufacturer.

Offset Current A compensating direct current injected into an amplifier input to assure zero output in the absence of an input signal.

Offset Voltage A compensating dc voltage applied to the input of a direct-coupled amplifier to assure a zero (or other reference) output voltage in the absence of an input signal.

Operational Amplifier (Op Amp) Very-high-gain, direct-coupled amplifier that uses external feedback to control its response characteristics. Can be a voltage-gain type (OVA) or transconductance-gain type (OTA).

OR Gate In a positive-logic system, a gate with more than one input whose output is a logic-1 if at least one input is also a logic-1, and whose output is a logic-0 only if all inputs are logic-0s. The reverse is true in a negative-logic system. Also known as Inclusive-OR gate.

Parity Bit A bit appended to an array of bits to make the sum of the bits always odd or always even.

Parity Check Test to determine whether the number of logic-1s or logic-0s in an array of bits is odd or even.

Positive Logic The reverse of negative logic; most systems use positive logic.

PROM/pROM Programmable Read-Only Memory. Read-only memory whose contents are programmed by storing a charge in a cell location through insertion of a large-voltage pulse.

Propagation Delay The difference in time between the application of a signal to a particular circuit or system and its appearance (or result) at the circuit or system output.

Pull-Down Resistor Generally, a resistor connected to ground or a negative voltage, e.g., from base to ground.

Pull-Up Active A transistor used in place of a pull-up resistor to effect low-output impedance without high power consumption.

Pull-Up Passive See *Pull-Up Resistor.*

Pull-Up Resistor Generally, a resistor connected to the positive supply voltage, e.g., from the collector to Vcc.

Radix A number arbitrarily selected as the fundamental number of a system of numbers, e.g., for binary numbers the radix is 2 (1 and 0).

RAM Random-Access Memory. A computer memory that can be reached (accessed) without using a specific address, and where data can be both stored (written in) and retrieved (read out).

RCTL Resistor-Capacitor-Transistor Logic. A form of DCTL wherein capacitors are put in parallel with the base resistors to improve circuit speed (reduce propagation delay). See *RTL.*

Read/Read Out Retrieve data from a memory.

Refresh Process whereby data in a memory is renewed (stored again) at periodic intervals.

Register Computer hardware used to store one machine word. Register length equals the number of bits it can store.

Reset Change the state of a flip-flop from "on" to "off."

Retrieve Same as Read.

Ripple Counter An asynchronously controlled counter; the clock (timing) is derived from an output of a previous stage.

Rise Time The time required for a transistor's collector current to rise from 10 percent to 90 percent of saturation; the time elapsing between the leading edge of a square-wave input to a circuit until the circuit reaches 90 percent of its steady-state value.

ROM Read-Only Memory. Memory containing prestored data which cannot be altered during operation; data can be retrieved but not written in.

RS Flip-Flop Reset-Set or Set-Reset Flip-Flop. Two-input flip-flop with the restriction that both inputs cannot be energized simultaneously because the resultant state will be indeterminate.

RTL Resistor-Transistor Logic Circuit. A direct-coupled transistor logic (DCTL) circuit where the inputs are coupled to the transistor bases through series-connected resistors.

Sample and Hold Circuit Circuit which samples an individual pulse and stores it for a subsequent operation; e.g., for comparison with another pulse.

Schmitt Trigger A flip-flop with one arbitrary input and one reference input whose output is a combination of both inputs; i.e., a voltage comparator.

Scratch Pad Small, fast memory used for temporary data storage, usually of interim calculation results.

Serial Mode Operation performed bit-by-bit, generally beginning with the least-significant digit. Also known as Series Operation.

Set Change the state of a flip-flop from "off" to "on."

Shift Register A storage register with the ability to shift its contents in one direction or the other (to a preceding or following register or other storage device).

Single Shot Circuit with one stable state and one quasi-stable state. When triggered, changes its state and falls back to the stable state after a time determined by its RC time constant. Also known as One Shot and Monostable Multivibrator.

Slew Rate The time rate of change of an amplifier output voltage for the largest input signal at which the amplifier performance remains linear; measured in volts per microsecond ($V/\mu s$).

Squelch Fault or noise suppression.

Storage Same as Memory.

Strobing Method of detecting the presence/absence of a signal via periodic insertion of a pulse into the circuit under test.

Synchronous Counter Counter controlled by an external clock pulse.

Threshold That value of signal voltage required to turn a quiescent device or circuit "on."

Threshold Logic Logic where the inputs are threshold functions (*combinations* of Boolean logic gates, i.e., combinations of ANDs, NORs, ORs, etc.).

Throughput The ratio of "good" blocks of data received to the total number of blocks transmitted, expressed as a percentage:

$$T = \frac{\text{good blocks}}{\text{total blocks}} \times 100$$

where a block of data is any preset number of bits, and a "good" block is one in which no bits are in error.

Toggle Same as Flip-Flop.

Totem-Pole An IC power-output circuit which can be connected directly to a load, eliminating the need for an output transformer.

Transparency The ability of a memory to read out data without destroying the data at readout; or to have new data stored (written in) without destroying data previously stored in the memory; or both.

Trigger (1) Same as Flip-Flop. (2) A signal input. (3) To start a device or circuit operating.

Tri-State Logic A trademark of National Semiconductor that connotes logic which has a "1" state, a "0" state, and a disable state that is a high-impedance condition allowing only a small leakage current (40 μA maximum) to flow into or out of the device.

Truth Table A table containing all possible input-logic combinations and the resulting output logic for these combinations.

TTL, T^2L Transistor-Transistor Logic. Logic circuits where the inputs are to the emitters of individual transistors that are connected as common-base amplifiers, and whose outputs feed the base of another transistor whose output is an inverted version of the input.

Turn-Off Time See *Fall Time,* last part.

Turn-On Time See *Rise Time,* last part.

Word A set of bits that is treated, stored, and transported (shifted) as a unit in a computer; the smallest addressable unit of information; the number of bits comprising it is the word "length"; can contain one or more bytes.

Wired-AND An AND function formed by *wiring* two or more outputs together.

Wired-OR An OR function formed by wiring together two or more outputs employing active pull-up and passive pull-down.

Write/Write In To deliver data into some form of storage.

Index